NEW YORK TIMES and
LOS ANGELES TIMES BESTSELLER

The Remarkable Love Story of
an Owl and His Girl

Wesley the Owl

"In between extraordinary owl tidbits, [O'Brien]'s . . . portrait of the complex and unforgettable animal she grew to love is irresistible." —*People* (★★★★)

STACEY O'BRIEN

Praise for *Wesley the Owl*

"Four stars [out of four] . . . fascinating . . . [O'Brien's] portrait of the complex and unforgettable animal she grew to love is irresistible."

—*People*

"Her heartwarming story is buttressed by lessons on owl folklore, temperament [playful and inquisitive], skills, and the brain structure that gives them some amazing abilities, like spotting a mouse under three feet of snow by homing in on just the heartbeat. This memoir will captivate animal lovers and . . . should hold special appeal for Harry Potter fans who've always envied the boy wizard his Hedwig."

—*Publishers Weekly*, starred review

"Sweet, quirky memoir. . . . [T]his little guy's such a character, you'll miss him, too."

—*USA Today*

"[Stacey] was able to observe owl behavior as no one else before her has been able to. . . . This book will be of special interest to lovers of nature, especially bird lovers. It will be of interest to readers who are just looking for a good read that tells the story of a young girl's love and devotion to animals, and that animal's love and devotion in return."

—Paul Pessolano, Borders Books & Music, Snellville, Georgia

"From Christopher Robin to Harry Potter, there's a valid reason why owls make great companions to humans: they've got something to teach us about ourselves. Filled with laugh-out-loud adventures, *Wesley the Owl* is definitely a tale for grown-ups but one which makes us appreciate what it is to be young again . . . the journey is remarkable and uplifting."

—Geoffrey Jennings, Rainy Day Books, Fairway, Kansas

"After I finished reading *Wesley the Owl* I wiped away the tears."

—Joyce Ripp, Administrator, Trade Show Manager, Northern California Independent Booksellers Association

"I fell in love with Wesley and Stacey too. This wonderful book made me feel so good that I am recommending it to everyone."

—Norman Goldman, Barnes & Noble, Granada Hills, California

"It's a wonderful book and reminds me of a childhood favorite, *That Quail Robert* by Margaret Stanger. I am looking forward to hand-selling *Wesley the Owl*."

—Brian Woodbury, The Toadstool Bookshop, Milford, New Hampshire

"This book will most likely be just as successful as *Marley and Me*."

—Barbara Tavres, Barnes & Noble, Oakland, California

"I finished it in a big old puddle last night. [I]t both brought back the loss of my much-loved pooch a couple of years ago and reminded me why I loved my little mutt lying with her chin on my lap as I read last night! Stacey O'Brien wrote a wonderful memoir!!"

—Kelly Norman, http://bananas4books.blogspot.com/

"Movingly details how her life was changed and rearranged by this animal's intelligence and sensitivity, and how the power of the human-animal relationship surprised even her. This touching memoir will expand your knowledge and challenge your thinking about interspecies communication and compassion . . . [and] highlights little-known facts that make wise and mysterious owls even more precious to us."

—www.dailycamera.com

"I LOVE *Wesley the Owl*! Not since Konrad Lorenz have I read such an honest, vivid, and revealing account of the rich and complex life of an individual bird. Stacey O'Brien has captured the essence of the soul of an unforgettable owl. Affectionate, quirky, joyous, and wise, Wesley shows us the Way of the Owl—the way to God and grace. This book is destined to become a classic, and will deepen importantly the way we understand birds."

—Sy Montgomery, author of *The Good Good Pig*

"*Wesley the Owl* is beautiful, funny, transcendental, fascinating, and powerful. I LOVED THIS BOOK!"

—Lynne Cox, author of *Grayson* and *Swimming to Antarctica*

"*Wesley the Owl* is a heart-wrenching and heartfelt story of the deep, reciprocal, and enduring emotional bonds that developed between Stacey O'Brien and her longtime friend. It's an inspiring example of how animals are able to reveal to us who they really are and who we really are, when we allow them to express themselves openly and safely. Profoundly passionate and personal, this remarkable book shows how we can all increase our compassion footprint in a human-dominated world. Read it and share widely. I sure will."

—Marc Bekoff, University of Colorado; author of *Wild Justice: The Moral Lives of Animals*

"This compelling story sheds a bright, shining light into the world of animal emotions and the powerful bonds forged between animals and humans. A heartfelt journey of life and love with one of nature's wild creatures, *Wesley the Owl* is a must-read story of faith, compassion, and selfless devotion."

—Jay Kopelman, author of *From Baghdad, With Love* and *From Baghdad to America*

"Most 'me and my bird' stories are mildly entertaining at best, but *Wesley the Owl* is a different animal altogether. Stacey O'Brien got to know this owl with a unique combination of deep scientific understanding and rare emotional intensity, and the result is stunning, unforgettable. Read this book and you will never see owls, or humans, in the same light again."

—Kenn Kaufman, author of *Kingbird Highway* and *Flights Against the Sunset*

"This fun book reminded me of *Marley & Me*, but with wings. Warm, weird, and wonderful, *Wesley the Owl* is proof that man's best friend sometimes has feathers."

—Mark Obmascik, author of *The Big Year*

"The best of love stories between two intelligent beings, told (by the human) with good humor and remarkable insights into the mind of an owl—I couldn't put it down."

—Donald Kroodsma, Ph.D., professor emeritus at University of Massachusetts, Amherst, and author of *The Singing Life of Birds*

"Wesley will make you wonder if owls are not at least as wise as humans and as capable of compassion. *Wesley the Owl* will stretch your notions about the limits of interspecies communication and love. It will entertain, delight and, finally, cause you to weep. Guaranteed."

—Sam Keen, author of *Sightings: Extraordinary Encounters with Ordinary Birds*

"Stacey O'Brien tells the intriguing story of how her life was changed and rearranged when she attempted to tame and raise Wesley—a barn owl. She shows us how she was ultimately repaid with his love and devotion, and given glimpses into the mind of an animal that has an unexpected ability to understand human language and to communicate. Fascinating!"

—Stanley Coren, psychologist and author of *How Dogs Think* and *Why Does My Dog Act That Way?*

"From wingtip to wingtip, *Wesley the Owl* will open your heart to feathered wisdom. And you will learn some astonishing facts about birds and other creatures from a compassionate scientist dedicated to saving lives."

—Don Höglund, DVM, author of *Nobody's Horses*

"With an eye for detail not often seen in books about animals Stacey O'Brien tells her fascinating story with great passion. This wonderful, enchanting book makes one understand that the bond we have with animals goes beyond the norm of dogs and cats and can cross over to all of nature. *Wesley the Owl* is a true testament that love comes in many shapes, sizes, fur and feathers."

—Randy Grim, author of *Miracle Dog*, subject of *The Man Who Talks to Dogs*

"Stacey O'Brien's relationship with an unreleaseable barn owl spanned almost two decades, and she tells their story of mutual devotion with an irresistible combination of empathy, humor, and keen observation. *Wesley the Owl* captivated me from the first page . . . it is a beautiful, inspiring book."

—Suzie Gilbert, wildlife rehabilitator and author of *Hawk Hill*

Wesley the Owl

Photo by Stacey O'Brien

Wesley the Owl

The Remarkable Love Story of an Owl and His Girl

Stacey O'Brien

ATRIA PAPERBACK

New York London Toronto Sydney New Delhi

ATRIA PAPERBACK

A Division of Simon & Schuster, Inc.
1230 Avenue of the Americas
New York, NY 10020

First Atria Paperback edition August 2013

Book design by Ellen R. Sasahara

Manufactured in the United States of America

20

The Library of Congress has cataloged the Free Press hardcover edition as follows:

O'Brien, Stacey.
Wesley the owl: the remarkable love story of an owl and his girl / Stacey O'Brien.
p. cm.
1. Barn owl—California, Southern—Anecdotes. 2. Barn owl—Behavior—California, Southern—Anecdotes. 3. Human-animal relationships—California, Southern—Anecdotes. 4. O'Brien, Stacey. 5. Women biologists—California, Southern—Biography. I. Title.
QL696 .S85027 2008
598.9'7—dc22 2008007985

ISBN 978-1-4165-5173-7
ISBN 978-1-4165-5177-5 (pbk)
ISBN 978-1-4165-7981-6 (ebook)

Dedicated to my parents,
Ann Baker Farris and Haskell Glenn O'Brien,
who gave me wings to fly
and
In loving memory of my grandmother
Agnes "Zimmie" O'Brien,
who rescued, raised, and loved a barn owl
long before I was born

Author's Note

In the interest of brevity, I have not listed every place that Wesley and I lived over a nineteen-year period, nor every roommate we had. I moved several times over the course of this memoir, but it would not serve the reader to have to wade through all of that. Also, I have changed a few names of those portrayed.

Since this story took place, federal laws protecting wild birds have become more stringent. It is against the law to keep any wild bird or indigenous wildlife without a permit. If you find a baby bird on the ground, *leave it alone*. The parent birds will often be around, know right where the baby is, and will continue to feed it. Before interfering, first observe the baby bird, keep away potential predators (cats/dogs), and intervene only after you have observed it for a couple of hours, determining clearly that the baby is not being tended to. If you can reach its nest, try putting it back in the nest. Then watch to see if the parents return within an hour. Parents will *not* reject a baby bird that's been touched by a human or animal. Birds cannot smell the way mammals can. If you can't replace the bird in its nest, take it to an animal rescue center; workers there will know of a certified wildlife rehabilitator who can take care of it.

There is much urgent work still to be done to protect animals from abuse, exploitation, and destruction of their environment. You can join an organization such as Defenders of Wildlife, The Jane Goodall Institute, the Audubon Society, and the ASPCA, to help in the battle to protect our fellow citizens of the earth. It is also extremely rewarding to volunteer at your local wildlife rescue and rehabilitation center, which is always looking for—and happy to train—willing helpers.

Contents

1

The Way of the Owl

ON A RAINY Valentine's Day morning in 1985, I fell in love with a four-day-old barn owl. I'd been working at Caltech (California Institute of Technology) for about a year when one of the scientists called me into his office. He mentioned that there was an owl with an injured wing, and said, "Stacey, he needs a permanent home."

The little owl was so tiny and helpless he couldn't even lift his head or keep himself warm. His eyes weren't open yet, and except for a tuft of white down feathers on his head and three rows of fluff along his back, his body was pink and naked. I was smitten beyond reason by his hopelessly goofy appearance. He was the most wonderful creature I'd ever seen, gorgeous in his helplessness. And, oh, was he uncoordinated. His long, lanky legs stuck out awkwardly, and his oversized talons erratically scratched anyone who held him. His scrawny body had two little nubs that would eventually become wings, and his ungainly pterodactyl-like head wobbled from side to side. It seemed as if he had been assembled from the flotsam and jetsam of many different creatures.

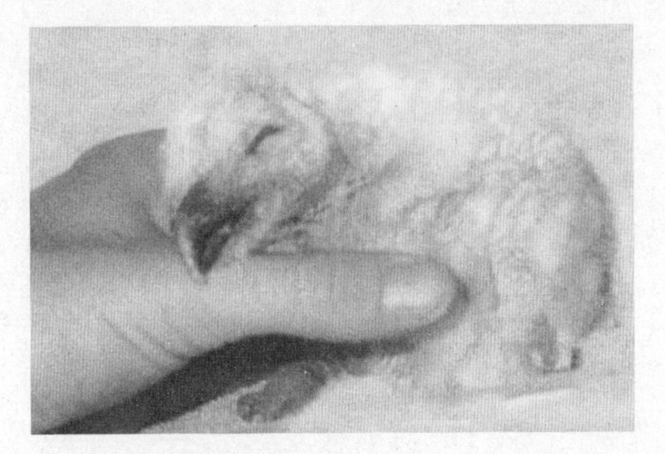

Wesley at four or five days old. *Stacey O'Brien*

Under normal circumstances, a rehabilitation center would have raised him using owl puppets to feed him and teach him to live in the wild, which is how biologists have raised endangered birds like sandhill cranes and the California condor that they intend to release. But this baby had nerve damage in one wing, so although he might one day be able to fly well enough to hunt sporadically, his wing could never build up to the level of endurance he would need to survive in the wild.

Like all barn owls, the baby smelled like maple syrup but not as sweet, something closer to butterscotch and comfy pillow all in one. Many biologists at Caltech, where I both worked and took classes, would bury their faces in their owls' necks to breathe in their delicate, sweet scent. It was intoxicating.

Scientists from all over the world were on our barn owl research team. There are seventeen species in the barn owl family and they live on every continent except Antarctica, but the ones we worked with were all *Tyto alba*, the only species that lives in North America. Found from British Columbia across

North America through the northeastern and southern United States, as well as in parts of South America and the Old World, barn owls are raven-sized birds, about 18 inches from head to tail. They weigh only about one pound full grown, but their wingspan is magnificent—averaging three feet, eight inches—almost four feet across. And barn owls are strikingly beautiful; their feathers are largely golden and white and their faces a startling white heart shape.

As gorgeous as they are, it is the owls' personalities that invariably capture the hearts of the people who work with them. All of the Caltech scientists grew intimately attached to their birds. One big, strapping scientist worked with an owl that got loose, flew into the ventilation system of the building, and there somehow hurt his foot. Owls are very sensitive and easily stressed. Even though the injury was minor and the owl was taken care of right away and not in any pain, he just turned his head to the side and wouldn't look at anyone or eat. Within a day, the owl died. The incident had so upset him that he turned his head away from life, and there was nothing any of us could do to coax him back. After he died, the big tough scientist sobbed and cradled the owl's body in his arms. Then he took a short leave of absence. That's how much the owls would work their way into our hearts.

This tragic behavior wasn't unusual for owls, who are emotionally delicate, even in the wild. For example, owls mate for life, and when an owl's mate dies, he doesn't necessarily go out and find another partner. Instead, he might turn his head to face the tree on which he's sitting and stare fixedly in a deep depression until he dies. Such profound grief is indicative of how passionately owls can feel and how devoted they are to their mates.

This is the Way of the Owl.

I LEARNED MY own passionate love of animals from my dad, who worked at Jet Propulsion Laboratories (JPL), one of Caltech's labs, for as long as I can remember. He would take my sister, Gloria, and me on many adventures to the ocean and into the Angeles Crest National Forest, which bordered our house. He taught us to observe animals without disturbing them, and every encounter was like a breathtaking meeting with an intelligent life form from another world, so different yet so familiar. I realized that each creature had its own personality. It was as rewarding for me to win an animal's trust as it would be for a space scientist to converse with an alien.

I learned to lure octopi out of their secret places by holding my hand very still, near shallow ocean rocks where they would hide. Because they are so curious, they'd eventually slide their tentacles toward me, gingerly explore my hands, then they'd gain confidence and end up crawling all over me. Gloria and I always tried to save baby birds we found, and we'd intervene to rescue lizards from cats. Once when I was four, my mother absentmindedly flicked a spider off a wall and flushed it down the toilet. I screamed and then cried for the rest of the day, because to me the spider was an innocent being who had hurt no one, and her life had been destroyed for no reason. My mother was flabbergasted by my extreme reaction and tried to reason with me, but even today I agonize about how casually our fellow creatures can be killed.

I also had an affinity for a more familiar, "traditional" animal. My first close bond, aside from my parents, was with our dog, Ludwig—half collie, half German shepherd. Luddie guarded my crib, lying under it while I slept and padding out to get my mother when I woke up. He always watched over me as I started to crawl and, when I began to learn to walk, would let me grab his tummy hair to pull myself up. I'd put my arms over his back, holding on to his fur, and he would walk very slowly and carefully with me. Whenever I started to fall, he went down, too, to cushion my landing. He taught me to walk, and I can still remember his soft, patient brown eyes looking back at me as we toddled along. I think Luddie's companionship lay the groundwork for my other relationships with animals, and I'm grateful that my mother had the wisdom to teach me to love and trust Luddie.

In addition to my love of furry, many-legged, complex animals, when I was a kid my bedroom was also filled with "experiments"—things in jars and vats of stagnant water full of exotic life forms that I could examine under my microscope. I once had two hundred silkworms in my room, which then hatched into two hundred mating moths that I would have to brush out of my bed at night before retiring. As a child I definitely had to clean my own room—and was instructed to use disinfectant. No one else would venture in there.

As I grew older, my father began taking my sister Gloria and me to lectures at Caltech, where I first saw my childhood hero, Jane Goodall. I was so convinced that I would grow up to do exactly what she did in Africa that I insisted on Swahili lessons. The next time she lectured at Caltech, I met her and tried out my newly learned language. I wonder if she remembers a little girl in blond braids who spoke Swahili.

Gloria and I were child actors who sang professionally in Hollywood recording studios into our twenties. We started singing onstage with our family band when I was five and she was three, and because we could sight-sing (read music and sing it without needing someone to teach it to us), a year later were doing TV commercials, movie scores, and singing background on albums. You probably heard us through the 1970s in campaigns for McDonald's and Pizza Hut, Little Friskies, ice creams, Bekins Moving and Storage, California Raisins, and more than a hundred other commercials. Gloria and I also sang backup for Glen Campbell, Barry Manilow, Helen Reddy, the Carpenters, and John Denver, among others, as well as in the second and fourth *Rocky* movies, *The Exorcist II: The Heretic*, and a bunch of Disney movies. Music definitely runs in our family. My grandfather was a drummer in the big-band era, and my dad's brother is Cubby O'Brien of the original Mouseketeers. Even so, because of my fascination with science and love for animals, it was more natural that I would go on to earn a degree in biology, which I did in 1985, at Occidental College—a sister school to Caltech, which had very few women at that time. Students at Occidental could enroll in any Caltech classes, and vice versa, which enabled both schools to broaden their curricula. I preferred the atmosphere at Caltech, though, so I took classes there, which also led to an undergrad, part-time job working with primates at their Institute of Behavioral Biology.

Eventually, I was offered another position in a department that studied owls. After monkeys, who were practically human, I was afraid that working with owls might be boring. Back then I ignorantly thought, as many people do, that they were "just birds." They seemed aloof, and I knew little about them beyond the fact that they flew around at night. How interesting could that be? The only owls I had ever been around were in my grandmother's massive owl figurine collection, and I figured real owls were probably not too different. But the owl job was full-time and I really needed the money. Plus it offered opportunity to participate in research. It was a plum position for a young biologist because I could learn so much from those around me.

I accepted the owl job, and within six months grew to love these emotional, sassy little creatures as much as did the distinguished scientists who had been working with them for years.

"STACEY," said Dr. Ronan Penfield, one of the scientists, "the zoos and other institutions are overwhelmed with owls that can't go back to the wild, and this owlet needs placement. Taking him home would be a perfect opportunity for you to do a long-term, deeper study of an owl on a level that's just not possible in a purely academic setting."

"You mean I should *adopt* him?"

"That's exactly what I mean. Since his eyes are still closed, he will imprint on you if you take him right away, and you could make observations, record his sounds and behaviors . . ."

I was both thrilled and terrified by this opportunity—scared by the enormous responsibility I would assume for this young life. I stared at Dr. Penfield to see if he was serious.

" . . . you may discover some things up close that we have not observed from a distance. I think this would be a beneficial study for our overall understanding of barn owls. You could keep us informed of your findings all along."

In spite of my fears, I wanted to leap across the desk, grab him by the shoulders, and shout, "*Yes*, I'll do it!" Instead I took a deep breath and tried to sound professional.

"I'll need to make some arrangements, but I would love to take him."

I was about to live with and raise one of the most beautiful animals on earth. Barn owls are quite different from all other owls. They are in a completely separate family called *Tytonidae*, while all other owls of the world are in the family *Strigidae*, meaning "typical owls." I was fascinated by all owls, but to have the chance to get to know the only representative of the nontypical owls that exist on the North American continent was very exciting.

The first bird, *Archaeopteryx,* began to appear in the fossil record during the Upper Jurassic period, around 155–150 million years ago. It had some dinosaur-like characteristics but was still clearly a bird. From then on, birds diverged, and owls would appear much later. My owlet was a bit of living history.

It's estimated that barn owls first started to appear in the fossil record during the Paleocene age (65–57.8 million years ago). The modern barn owl, *Tyto,* appeared around the middle of the Miocene period (23.7–5.3 million years ago) and diversified into various species during the Pliocene (5.3–1.6 million years ago) and Pleistocene (1.6–0.01 million years ago) periods. Wesley's species, *Tyto alba,* started showing up in the fossil record during the Pleistocene. Although owls are sometimes included in discussions of raptors, the truth is they are thought to be more closely related to nightjars than to diurnal (daytime) birds of prey (*Falconiformes*). Nightjars, which include whippoorwills, actually look like some kind of missing link between a regular bird and an owl.

Before I worked with owls I had never even heard of nightjars, and I used to skip over the parts of books that discussed how many millions of years ago an animal appeared on the world stage. But once I had my owl, this information became fascinating to me. His "tribe" had been here, probably living very close to where we were at that moment, for some 1.6 million years. What really blew my mind was that, in all that time, every single one of his ancestors had successfully bred and had a baby survive to breed. For 1.6 million years. There wasn't a single break in the chain, or he wouldn't have been here. Of course, this is true for every one of us who is on the planet—which seems like an incredible miracle.

Some scientists think birds may have evolved from dinosaurs, and to look at my owlet's feet and beak, especially before he had feathers, it sure seemed possible. Recent fossil discoveries suggest that some dinosaurs were warm blooded, had feathers, and kept their babies in nests, feeding and caring for them just like the parents of birds do now.

Another attribute that makes owls unique is their brain structure, which is completely different from that of most vertebrates. The barn owl's cortex is mostly dedicated to processing sound rather than visual images. I wondered how that would affect the way the owl interacted with me and my visually oriented domestic world. He must have a very different viewpoint, foreign to us. His world would be even more different from, say, a dog's, because dogs process their sensory information primarily through their noses and eyes. Dogs are mammals and social, so we humans and they have learned how to get along and live with each other over millennia. Some scientists even think that dogs and people helped each other evolve to our current forms. But it would be challenging to learn to live with this nonsocial animal. Owls don't stay in flocks, but individuals are devoted to their mates, living a mostly solitary life together.

Not only are owls interesting creatures historically and physiologically, but their temperament is also unique. Owls are playful and inquisitive. A friend of mine knew someone who had rescued a little screech owl and she described it as acting like a kitten with wings. She said the owl would fly up, then pounce on all kinds of objects exactly as a kitten does. Owls could also be creative. Sometimes I'd be walking by an office in the Caltech Owl Lab and see an owl making up his own game—throwing a pencil off a desk just to watch it fall and roll on the floor, then flying off the desk himself, twisting in the air to get a good angle, then pouncing on the pencil. I also saw postdocs talking nose-to-beak with their owls when they thought no one was looking; rubbing noses, kissing, and playing little games. They seemed to enjoy each other's company in the same way that dogs and people do. Could my owl and I develop such a great rapport? I wanted to find out. After all, this curiosity and desire to experience animals and learn from them directly is what drives a person to become a biologist or naturalist in the first place, much as the space scientist is driven to find out what's on that next planet or in that new star system. Perhaps this was finally my chance to get to know a wild animal in the way I had always dreamed of as a child. I wouldn't be traveling thousands of miles and bushwhacking into the jungles of Africa or the Amazon to find my special animal. My owlet was coming to me.

2

To That Which You Tame, You Owe Your Life

AT THE TIME I adopted my owlet, I was renting a room from my best friend, Wendy, who lived with her husband in a ranch-style home in a Southern California area where everyone had horses and other farm animals. In Wendy's case this included chickens, a flock of noisy geese, an Andalusian stallion, and a goat and mare that were inseparable. Wendy was a painter and musician. She and her husband were frequently on the road performing in concerts, and I looked after all the animals when they were away. Wendy was also pregnant and was going to need my extra help with the baby as well.

We'd been friends since I had somehow bested her at a rowdy game of flag tag on horseback at a summer camp in the San Bernardino Mountains where Wendy taught horseback riding. Wendy had a special way with horses, and over the years I honed some of my techniques for working with wild animals by watching her. Although I was twelve and she was eighteen when we met, the age difference didn't seem to matter because we shared so many of the same interests, including a passion for music.

When I first mentioned the injured orphan owl to Wendy, she was out in the barn taking pictures of her mare. I was a little nervous about asking if I could bring an owl home to live with us, but she just smiled and patted her horse on the neck.

"A barn owl! Won't he be a great addition to the family!"

"Wendy," I said, "I'll have to keep dead mice in the freezer and cut-up mice in the fridge. Would that be okay?"

"Meat is meat," she shrugged.

"I mean, a lot of mice, Wendy."

"A lot of mice? How many is a lot?"

"Well, probably thousands of mice as time goes by."

"How long do barn owls live, anyway?" she asked.

"I'm not sure how long they live in captivity. Maybe up to fifteen or twenty years."

"Well, I think you should do it," she said. "It's the chance of a lifetime."

IN THE WILD, the father owl hunts relentlessly. He has to provide approximately six mice per baby per night. The usual brood is five babies. The father also has to feed his mate, who never leaves the nest and eats about three mice per day. And he must feed himself about four mice a day. This adds up to some thirty-seven full-grown mice every night during nesting season.

A father owl is constantly harassed and henpecked, as the screeching and begging sounds from the nest are never out of his earshot. Wild males hunt like crazy, continually racing back to the nest where unruly babies all mob him at once, demanding food, virtually attacking the beak that feeds them. When the babies are older, the father avoids this confrontation by swooping in, hovering above the nest, and dropping his mouse payload from a safe distance. Then he zooms off to do it again.

In the wild, barn owls do not live that long. Only one out of fifteen even lives through the first year. They get hit by cars, since they tend to fly at about the same height as a car or truck and are confused by the lights and noise of traffic. They die by flying into live electrical wires, and they're especially vulnerable to poisoned meat, such as a mouse that's eaten poison. Habitat loss is probably the main cause of drops in barn owl populations. Domestic cats, allowed to roam outside, occasionally kill barn owl babies—as do raccoons, bobcats, mountain lions, coyotes, and bigger species of owls, such as barred and great horned owls, harriers, hawks, and eagles. Barn owls may even get mobbed and pecked to death by smaller birds like crows and ravens. Worst of all, some people shoot them "for fun" or because they dislike birds of prey, though it's a federal crime punishable by a mandatory jail sentence and a large fine.

The one out of fifteen barn owls that makes it into his second year faces all of these dangers. In the wild, the females lay one egg a day for about five days. When the eggs hatch, one per day, each owl is one day older—and bigger—than the next. Since the babies grow so fast, the first will always be the biggest and strongest; the second baby will be the next biggest and strongest, and so on down to the smallest and weakest last-born. The parents feed the strongest, most aggressive babies first, until they're completely full. In a good year everyone gets fed, but in a lean year, the youngest babies starve. As cruel as this may sound, it's better than underfeeding all of them and having them all die.

Of course, some weak, "unfit," traumatized owls, like my orphan baby who survived long enough for a kindhearted human to find and rescue him, are brought to rehabilitation centers.

At Caltech many of our tame barn owls would attach to one scientist and usually hang out in that person's office, though truth be told, they had the run of the whole floor. Other wild barn owls lived together in aviaries and flight runs in a huge barnlike building near our offices. I worked with a team of enthusiastic scientists doing top-notch research in a fun, easy-going atmosphere. We were encouraged to observe our birds as closely as possible in order to gain a deeper understanding of them. Dr. Penfield had kept his own owl at both his home and office for nearly two decades.

Because owls have no external genitalia, it's nearly impossible to distinguish males from females, even for an expert, unless you were to perform a minor surgery that allows you to see their organs. Dr. Penfield hadn't wanted his owl to be surgically invaded, and neither did I, so we both opted to guess at their sexes. We both guessed male, but to Dr. Penfield's chagrin, at around age fifteen his owl laid an egg.

Owls fly fast, even many that are not releasable into the wild, and it wasn't unusual to round a corner in the hallway at our lab and have to jump out of the way of an owl flying at you full steam. Our big, open conference area had floor-to-ceiling windows, and if we forgot to close the curtains on a beautiful California day, a flying owl might *wump* into the glass. Fortunately, owls are extremely well padded, with many layers of downy fluff and feathers, so they weren't injured or killed by these mishaps as other birds can be. But the entire length of the windows had perfect cartoonlike owl-shaped imprints where bits of their feathers stuck to the glass.

Tame barn owls were part of our daily lives, wandering through our offices, sometimes walking with their galumphing gait, sometimes flying like fighter pilots, sometimes hovering, sometimes gliding. It was a lot like working at Hogwarts, except that, instead of receiving letters from our owls, we would find coughed-up owl pellets in our coffee mugs. Of course, owls on the loose seemed strange to the uninitiated. One day an electrician came to work on the building's power supply, when, seemingly out of nowhere, an owl flew around a corner right at him. The poor guy let out an unearthly scream and hit the floor, covering his head and yelling in Spanish. In Mexico, it is believed that an owl's appearance can predict a death: owls are bad omens there and supernatural powers are attributed to them. I ran to the electrician, knelt on the floor, and explained the owl's presence in Spanish as he peered at me through his arms, cowering in fear. He didn't seem to believe me because he got up and raced to the nearest exit.

More in line with the Native Americans of Southern California who believe that owls are sent to guide us through dark places as friends, I consider owls a good omen. Certainly barn owls have been friends of farmers for thousands of years, keeping down the mouse populations in barns and grain storage areas. Barn owls might be the source of other myths, though, like the Celtic banshee—a spirit who screams at night—and the belief in haunted houses. The screech of a barn owl as it rockets out of an abandoned house can be terrifying if you don't know what it is. Today, we might liken it to the sound of an eighteen-wheeler losing its brakes. Imagine hearing that sound before the Industrial Revolution! Surely only a demon could have made such a harrowing noise.

But, for me, every time I have been about to go through a major change in my life—for the better—a flesh-and-blood owl appears right at the time I'm making a decision. They do seem to appear to guide me. One Sunday evening, returning from a weekend in the Sierra Mountains, I had the rural road to myself and the driver's side window down to enjoy the breeze. Out of the blue, a barn owl swooped down to eye level—right alongside the car. He flew next to me, almost touching my shoulder with his wing. I could have easily reached out to touch him. I was coasting down a hill and going slowly enough that I could look over at him—our eyes met as he looked over at me!—then we both looked back to where we were headed; then back at each other. We kept up this back and forth till we came to a curve in the road, at which he peeled off into a meadow. It was an amazing close encounter. (And it made me feel confident about what I had to do when I got back home.)

When I began working at the institute's owl aviary, among the barn owls were also a few great horned owls, some burrowing owls, and several long-eared owls with nowhere to go. We took in these other species out of a sense of charity, rather than for study. Unlike the tame owls that stayed with us in our personal quarters, all the wild owls were occasionally fed rats in addition to mice, because of their easy availability. Mice are about the size of a Snickers bar, and owls usually swallow them whole with nothing left over, which is much cleaner. But mice were more expensive and harder to come by. Hence the rats.

Some research lab from which we got our rats must have been doing genetic experiments, because their rats were huge, at least two feet long. They were delivered whole and frozen, and we used a meat cleaver to chop them into slices, like frozen rat pucks. We'd place the thawed, fur-edged rat patties into the aviary runs, where the owls would pick at the meat. Stray meat would fall through the slats of the aviary's wooden floor and onto an undulating layer of cockroaches three feet below. They made quick work of the rat meat, but they also had the annoying tendency to come up from below and crawl around in full view. The owls themselves lived about twenty to thirty feet above all this.

Because we did not want to enter the areas that held the wilder, more nervous owls too often, and overstress them, we hosed the remnants of the discarded and rotting rat parts daily to the back of the aviaries, where they developed into putrid piles. The stench was off the charts.

There's a definite cultural pressure among biologists to withstand the extremely gross without reacting. It has to do with gaining credibility among other scientists. Part of my job was to feed and clean up after the animals, so about every three months I'd go into the owl aviaries covered from head to toe in rain gear, heavy boots, a helmet, gloves, and face shield. I'd shovel the piles of green rat sludge, which writhed with zillions of maggots, into large garbage bags and hauled the whole mess into Dumpsters. *Ugh.* Maggots rained down on me with a pattering sound. The smell was like a living thing that swirled around me and beat its way into every nasal membrane and pore in my body. As a matter of survival we are programmed to flee from death and decay; I did want to run, but I had to override instinct to get the job done. I was quite proud of myself for not passing out.

For obvious reasons this feeding system wasn't too great, so we eventually constructed a multimillion-dollar, owl-friendly system of aviaries—blessedly designed with a self-flushing subfloor—in another building. When the day came to transfer the owls from the old to the new building, we caught them with special flex nets and put them in cardboard cat carriers for the short trip. Owls have strong opinions, which they are not afraid to express, and they were outraged, screaming all the way to the new building until they were freed into their facilities. Few sounds are as hair-raising as an enraged, screaming barn owl, and because they were in cat carriers, rumors started that we were in the business of torturing cats. Honestly, if we had been torturing cats, would we have been carrying them around the campus, screaming, in boxes with big drawings of cats on the sides?

The complicated job of maintaining so many owls made the prospect of the daily care of one owlet seem pretty simple. I would not have to deal with massive aviaries with water systems running under their floors, controlled climates, and daylight periods. I would need only a few items, like a good owl perch; owls cannot be kept in cages, as they tend to break their wings on them. My owl would live right with me in my bedroom—along with my collection of zebra finches—and I would just clean up after him every day.

The night after I agreed to take the owl, I didn't sleep well at all, thinking about the little soul whose life would become intertwined with mine. Taking him in would make it impossible for him ever to go back to the lab's aviary with the other owls, or to any other facility, because he would bond with me. It would be up to me to give him a happy, comfortable life. He would be completely dependent upon me physically and emotionally, and if I were ever to abandon him, I could doom him to die from fear, confusion, and grief.

I don't even remember the drive to Caltech the next morning. I raced up to the office to say I could take the baby home with me. The owl was in an incubator, and I reached in for the naked, helpless little creature. When I held him in the palm of my hand, I'd never felt so protective of anything in my life.

Seeing me with the owl, Dr. Penfield smiled.

"To that which you tame, you owe your life," he said.

3

Owl Infancy

I NAMED THE owlet Wesley. The name seemed perfect; cute enough for a baby owl yet wise and sophisticated for when he grew into a distinguished adult bird of prey. I had met him on Valentine's Day, and eventually his adult face would be shaped just like a white heart, so Valentine became his middle name. I cupped the tiny owlet in my hands, held him against my cheek, and said, "I am your mother now." I tucked him into the deep pocket of my lab coat, gently placed my hand around his frail body to keep him warm, and carried him that way from building to building as I gathered material to make a nest box, including receiving blankets I had brought from home to simulate a nest.

Wesley went everywhere with me from then on. I even wrapped him in baby blankets and held him in my arms while grocery shopping, to keep him warm during that first cold winter. Occasionally someone would ask to see "the baby" and when I opened the blanket, would leap back shrieking, "What is that?! A dinosaur?" Apparently, the world is full of educated adults with mortgages and stock portfolios who think that people are walking around grocery stores with dinosaurs in their arms.

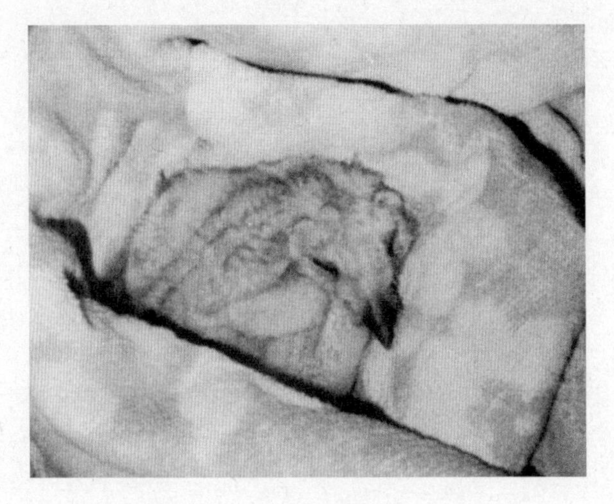

About one week old. *Stacey O'Brien.*

At Caltech, the scientists had discovered that owls hunt by the sound of their prey—not with echolocation like bats, but from homing in on the tiny noises that the prey animals make, triangulating the mouse's location with their ears. When we humans enter a forest, we primarily use our sight to navigate through the trees around us. By contrast, even though owls' night vision is keen, their primary sense is auditory. Every barn owl has a "sound picture" of the cacophony of noises within the forest, including those made by animals, leaves blowing across the ground, and wind moving through the trees. A barn owl's satellite dish–shaped face focuses and receives the sounds, directing them to its ears. Unlike human ears, which are in the same place on each side of the head, owls' ears are irregularly placed. One ear is high up on the head and the other is lower, so that the owl can triangulate the location of a sound much more accurately than a human can. The owl brain's large cortex is dedicated to auditory processing in much the way that ours has evolved for visual mapping, so it creates an auditory map of his world. As a result, a barn owl can accurately locate a mouse under three feet of snow by homing in on only the heartbeat, and can hear its footsteps from extremely far away.

Knowing all this, I talked to Wesley constantly while his eyes were still shut so that, when his eyes opened, he would have already bonded to my voice. The same dynamic would occur in the nest, where the baby owl would hear his parents communicating with each other while he was in the egg. When Wesley opened his eyes for the first time, he stared right at me.

"Hello, Wesley," I said.

"Screeeeech," he hoarsely and softly replied, gazing deeply into my eyes.

Like all barn owls, Wesley had two sets of eyelids: the nictitating membrane, which was sky blue, underneath his regular eyelid, which was pink and had perfect white lashes, which are actually tiny feathers. If I had been a real owl mother, this interaction would have been similar. In the nest, owl mothers twitter and chirp to their babies even while the owlets are still in the egg, shortly before they hatch. When the babies' eyes open, they make eye contact with their parents and siblings as they chirrup to each other. The owlets are especially intense with eye contact when they are begging for food or lobbying for their mother's attention. Wesley focused on me right away, twittering and chattering, looking me in the eyes and trying to communicate. I was astonished at the intensity and clarity of his focus on me.

Wesley's eyes were a deep, inscrutable black. Even when they first opened, they harbored a great mystery and held my gaze. Looking into his eyes was like looking into infinity, into something far away and cosmic. It was a profoundly spiritual experience—I never tired of it—and I was often startled by his eyes' wildness and depth. Many other people noticed this quality throughout Wesley's life and struggled to describe the strong conscious personality that they detected behind those eyes.

As with all owls, Wesley's eyes were fixed in their sockets, so the only way he could get depth perception was to move his head from side to side. He also had an extremely long, thin neck under all that white down, and could contort his head in dozens of different ways, including turning it a little more than 180 degrees—that disconcerting, spooky habit for which owls are known. In the wild, owls can sit comfortably with their heads facing backwards, watching for both predators and prey without so much as lifting a foot. When he was still white and downy, Wesley would sometimes freak me out when I'd realize that he was facing me over his back. Even though it's completely natural, it seems unnatural, even supernatural, like a scene from *The Exorcist*. "Wesley," I'd tell him, "don't scare me like that!"

Wesley observed the cardinal owl rule of never pooping in his nest. From the beginning, owls are very persnickety about hygiene, and Wesley scooted backwards until his bottom was as far as possible over the edge of the "nest" I'd made for him before pooping. When he first started to scoot around on the carpet, he would back up with his rear end high in the air and push backwards, trying to find the end of the rug so he could poop. He'd go quite a distance trying to find a suitable spot. Because he seemed to think of the carpet as nesting material, I realized that if I lay down a paper towel behind him, he would notice the change in texture and decide that he had reached the end of the "nest" and would poop there with a quiet air of dignified relief.

When describing both the act of defecating and the substance of fecal matter itself, biologists prefer to use the scientific term "poop." It's both a noun and a verb. A popular field of biology called scatology is the study of scat, which is not to be confused with mere poop. Although technically they're the same, we call it "scat" if we are studying it to learn something about the health and diet of an animal. When the animal has pooped on us or has ruined something with his pooping, we tend to use the term "shit," as in, "Oh, man, he just shit down the back of my neck." So if it's on the ground, it's poop. If it's under your microscope, it's scat. If it's running down your neck, it's shit.

Wesley not only produced copious amounts of hot, slimy poop and a fair amount of shit, he also had very acidic saliva, which burned a little when he kissed me on the cheek. That saliva is the first step in an extraordinary system that allows owls to digest an entire mouse in an hour. It also fights bacteria that owls might encounter when eating meat that is slightly "off." Once they've digested the mouse in their two-part stomach, they cough up a pellet of hair and bones with all the meat completely stripped off. Pellets are now in high demand for school and college biology classes because each one contains a complete rodent skeleton. Dissecting pellets introduces students to one of the ways biologists learn about the habits and diets of wild animals. I wish I had thought of this before throwing out thousands of Wesley's pellets. Today, however, someone does sell the plentiful leftovers from owl aviaries everywhere. Acid bathed in the stomach, the pellets are actually pretty clean, unlike a cat's hairballs, and they dry and harden quickly. As time goes by, they disintegrate, and if left in the nest they can become a soft, fluffy layer at the bottom of the nest. A lot of people mistake pellets for poop, but the owl's digestive system sorts it all out very efficiently. Owl poop looks just like other bird poop.

Unlike other owls, barn owls have a defense mechanism much like that of a skunk. When threatened or extremely stressed they shoot out, from their back ends, a greasy dark brown stuff that smells horrendous. It's not from a special gland, as is a skunk's spray, but it's a great deterrent nonetheless. Wesley never did this to me, but throughout his life he would emit this substance shortly after a stressful incident. I usually had enough warning to catch the dark goo on a towel and race it out of the house and into a trash bag, lest the entire place have to be evacuated. Once in a while Wesley sprayed for no apparent reason, so it may also be a way to excrete toxins.

Wendy's baby was born soon after Wesley arrived, so they were about the same age. We joked that my baby, Wesley, was a lot more trouble than her baby, Annie, because Annie slept through the night, whereas Wesley didn't until he was several months old. Though owls are nocturnal, Wesley eventually learned to sleep or keep himself quietly entertained throughout most of the night, trying to do exactly what I—his mother—did.

Every night I put Wesley's little nest box next to my pillow, secured on the bed so it wouldn't fall. In the wild a baby owl is never separated from his mother, so I'd sleep with my hand draped over him in the box. In the nest, mothers and babies cuddle together and the mothers gently preen the babies with soft nibbling motions that are very comforting to the owlets, so I did this for Wesley with the tips of my fingers acting like a mother's beak, and Wesley responded by nibbling my fingers and getting sleepy. He snuggled against my hand, sleeping on his stomach, tucking his head and legs under his tummy and becoming a little owl ball. We'd sleep peacefully for about two hours and then I would awaken to the most urgent sound—a cross between a screech and a hiss—which meant Wesley was hungry.

After bringing Wesley home, I had set up a little feeding area at my desk with a towel for him to sit on. I'd get up, grab a baggy of mice out of the freezer, and defrost them in the microwave. Then I used scissors to cut the mice into owlet-bite-sized pieces. After setting Wesley on the towel, I'd support him with my left hand while feeding him with my right. I tried to emulate an owl parent by using tweezers to rub the little mouse chunks against the side of his beak. He'd grab the morsel sideways and wolf it down. Wesley would eat until he was so full that he would look nauseated and turn away his head in seeming disgust when I offered another piece. Eyes drooping shut, head held to the side, he'd then fall asleep against my hand. It was the beginning of a lifelong tradition of cuddling.

Wesley was growing so quickly that every morning I could see that he was bigger than the night before. He needed mice to keep up with his growth. Lots of mice—six per day. But I had that under control. Caltech was providing me with large bags of neat and tidy prekilled, frozen mice. Mouse McNuggets.

Then out of the blue there was a crisis—a statewide shortage of rodents. It was every predator and snake owner for himself, and the lab politely told me to go kill my own mice. Even though Wesley was a bird of prey, it had never occurred to me that I would actually have to kill animals to take care of him. I was horrified, but there was no other way. A mother will do anything for her child, and if her child can only eat mice and will die without mice, then a mother will kill mice. And so it began.

At first I didn't know how to kill mice and I wanted it to be as quick and painless as possible. I couldn't use chemicals, so, on the advice of fellow scientists, I set myself up with a bag of mice in the backyard and cut off their heads with sharp scis-

At about five weeks in his nest box. *Stacey O'Brien.*

sors, which was so fast the mice never knew what hit them. But it was grisly, and afterward the backyard resembled a mass murder site.

The carnage attracted quite a number of cats, which would walk into the yard single file and line up about two feet away, watching silently, their eyes fixed on my every movement, just the tips of their tails twitching. They would look solemnly at the jostling bag of live mice, then at me cutting off their heads, then at the twitching bodies on the grass. I would reach into the bag, grab another mouse, and do it at least thirty times over. The cats never approached the bag of mice or me. It was eerie. I felt like a high priestess performing a ritual sacrifice for worshipers.

After going through all the mice in a bag, I'd shovel the mouse heads and bodies into a plastic bag and put them in the freezer. This became a problem, because when it was time for

Wesley to eat, the blood had congealed the remains into frozen clumps, so I had to chisel pieces off one by one with an ice pick. It was messy work and I started having bad dreams.

Eventually, I had to find another solution. Someone told me that if you held a mouse by its tail and swung your arm back like a baseball pitcher and slammed it into a hard surface flicking your wrist at the last moment, it would die instantly, painlessly, and unconscious of even being threatened. I would not need to try twice. So I began using this technique and it worked perfectly and painlessly for the mice. I, however, developed problems with my right wrist—carpal tunnel syndrome—from killing about 28,000 of them throughout Wesley's lifetime (at more than a dollar a mouse).

To find that many mice was no easy task. I staked out every pet store within a twenty-mile radius of my home. I had to have mice. Whole mice. Each organ, each bone, each hair on its body was crucial to Wesley's nutrition. Some people have tried to feed owls strips of meat rolled in calcium, but without every bit of the mouse carcass, barn owls die a slow death of blood poisoning. They are perfectly adapted to eat whole mice. So I committed to memory the stores' delivery schedules and traveled as far as I needed to get my quota.

On one particular night, I went farther from home than usual to find about thirty mice at one store—I bought every mouse they had. They weren't the usual white ones, but were brown like local field mice. Fortunately, color made no difference in nutritional value. I needed them all and needed to get them home fast. The pet store owner stuffed them all in a paper bag, which I put into the backseat, and got in my car. On my way home I noticed that I was low on gas and pulled into a full service station.

As the attendant walked up to the car, I rolled down my window and said, "Good evening." He nodded and then glanced into the backseat of the car, started, looked back at me, then again to the backseat. *Oh, no,* I thought.

"Is anything wrong?" I asked the attendant.

He didn't answer, just slowly backed away from the car. But I knew what was wrong without looking behind me. The mice were loose.

During transport, the mice often ate their way out of the paper bags. It wouldn't occur to me until months after adopting Wesley to use an aquarium in the backseat to store mice. Sometimes I didn't catch all of them, either. They'd get stuck behind the dashboard and eat through the wires to my radio or electronic signals, which ruined them. I could not afford expensive repairs. One mouse died in the dashboard beyond reach and rotted. For months I had to drive with my windows open to air out the odor.

Another mouse fell out of the dash and into my open shoe one day. I screamed and briefly lost control of the car, then pulled to the shoulder, pushed open the door, and yanked off my shoe. Out fell a struggling, grease-covered white mouse that quickly escaped into the bushes.

Back at the gas station, as the attendant continued to back away from the car, I asked, "Um, can you fill it with regular, please?"

"Lady," he sputtered, "your car is completely infested with field mice. I mean . . . the entire inside of your car is crawling with them!"

I finally looked into the backseat, where thirty brown mice were running all over the seats, the floors, the handles, the windowsills . . . everywhere. It could've been a scene in *Willard*. I glanced up, but the attendant was gone.

It would be another self-serve night.

4

Barn Owl Toddler: Love Me, Love My Owl

About two months old and curious about the camera.
Wendy Francisco.

I COMMUTED TO work every day and brought Wesley with me. It was sometimes difficult to do my job while carrying him around, so one day I tried leaving him with one of the researchers.

"Hey Jergen, would you babysit my owl for a while?"

Jergen looked into the box, saw the sleeping white ball, and said, "Sure thing, yah!"

Wes was sound asleep when I left him, and I rushed through

my work hoping he'd sleep long enough for me to get a lot done. But minutes later, I heard boots clumping across the big wooden floor in the owl barn and there was Jergen looking even paler than usual with a bright red blotch on each cheek.

"You've got to get back up to the office," he panted.

"What's wrong?" I cried, dropping everything. "Is he hurt?"

"I don't know, I don't know, just go up," he said.

With my heart thundering in my chest I raced back up to the offices, taking the stairs two at a time. As I got to the third floor I could hear a horrible commotion of screeches. I ran into the room, and there was Wesley's little white head bobbing up and down over the top of his nesting box. He was screaming so loudly that he had cleared out the room.

When I got to his box he just lowered his eyelids in greeting and gave a soft, sweet little twitter. All was well with the world now that I was with him again.

"You can't leave him up here anymore," Jergen said. "We'll never get anything done."

In the wild, there is no such thing as a babysitter. It's the Way of the Owl.

After this scare, I took Wesley absolutely everywhere with me and did not leave him alone again until he was three months old, at which age he would have been starting to leave the nest. But until that time, I carried him in his box into every room as I worked. He stayed in his box by my side while I was feeding other animals and caring for them and whenever I was in the lab working with microscopes and other instruments.

LIFE WAS MUCH easier at home because I could give almost all my time and attention to Wesley. And Wesley observed me carefully. Normally he would have taken all of his cues from his mother. Now he was taking all of his cues from me. So when he saw me petting Courtney, Wendy's golden retriever, Wesley was unafraid and curious. Being a baby meant he didn't yet know his wildness and was open to making friends with any animal that came along. If his mother thought someone was okay, so did he. Courtney had been curious about Wesley for quite a while, so I decided to let them meet. Wendy supervised Courtney, and I held Wesley. They touched nose to beak and neither of them reacted. I set Wes on the ground and Courtney sniffed him over, then lay down next to Wesley as if he were her puppy. From then on, Wes felt right at home sitting between her front paws, and they would just hang out together.

With Courtney the dog. *Stacey O'Brien.*

Whenever I took care of Wendy's daughter, Annie, who was born shortly after Wesley came home with me, I'd always involve Wesley, too. I could also leave Wesley in his nesting box next to my pillow in the bedroom, where he was used to sleeping, while I ate dinner with Wendy and her family. But the first three months seemed much longer than they really were because toting a fragile baby owl around was complicated and inconvenient.

One night when I was home with Wes, who was about a month old, the phone rang. I answered, and a soft, low voice said, "Hello, Stacey?" I went weak in the knees. The guy I had hoped for and schemed over, prayed for and cried over, Paul, was asking me out on a date. He was gorgeous, a musician, blond like me, short like me, into music like me. I had been pretty sure for a long time that Paul would make the perfect husband—it was just a matter of him realizing it. Did I mention he was gorgeous?

"I would love to," I sputtered.

But then with a shock I remembered the Way of the Owl—and the "no babysitter" rule. I would have to bring Wesley with me. I could just imagine what would happen if he started screaming in a fancy restaurant. I had to say something so I blurted out, "Um, I'm raising a baby owl and I can't leave him. Can I bring him with me in his little nest box with his little bowl of food? He eats mice, uh, but it's okay because they're already cut up." I was dying a thousand deaths. He hesitated for a moment then said, "Sure, no problem." We set up a date for two weeks later and hung up. I was thrilled.

Wesley was growing quickly. As he grew, he became more active, all legs and talons, pulling himself out of his blankets and trying to climb all over me. One night, as I was feeding him his cut-up mice, he suddenly lurched to his feet for the first time. He was six weeks old; he seemed surprised to be so tall and looked to me for an explanation. "Wesley! You're standing up all by yourself!" I told him. He seemed reassured.

Wesley at five weeks old, with Stacey.
Wendy Francisco.

At about seven weeks, standing next to a standard paper towel roll so you can see how big he is.
Stacey O'Brien.

Shortly afterward, Wesley took his first step—and with that step everything changed. He was like a human toddler, bounding around getting into everything in the bedroom. How was I going to protect an owl toddler? Fortunately, Caltech had loaned me a perch designed specifically for owls at this transitional stage. It was essentially a low mock tree stump nailed firmly to a 4-by-4-foot platform that sat on the floor. Wesley would be tethered to the stump with a leg jess. Next to his new perching post, I set his old nest box on its side, where he continued to sleep and groom himself.

Now that Wesley was mobile, I needed to introduce him to the leg jess and figure out how to leash him to the perching post, since I couldn't corral him twenty-four hours a day. Most falconers put a leather jess on both legs of their bird, which they attach to a leash whenever taking the raptor outside. I modified this design and put a very soft leather jess on only one ankle—so Wesley would get the idea but wouldn't have his legs tied together when on his leash. The leash was attached to the top of the perch-stump. He seemed to enjoy the new setup and the freedom it gave him to jump down and walk around on the floor. I leashed him whenever I couldn't be there to supervise him, as he didn't know what was dangerous and what was okay to play on, and I didn't know what kind of trouble he could get into without my watching him. At this age, the babies would still be with their parents in the wild, learning to do what their parents did. They would be climbing out of the nest and sitting on tree branches, but still very dependent on Mom and Dad for guidance and food.

Once Wesley could walk confidently, he waddled behind me from room to room. Off his leash, he followed me everywhere and watched everything I did. And everywhere he went, he hurried. He'd put his wings up as if flying, although it looked as if he were playing "airplane." He would bring each foot way up to his chest, then shoot that foot out as far as it would go, throwing his weight forward onto it, while pulling the other foot all the way up to his chest to do the same. The result was a hilarious galumphing gait that he kept for the rest of his life. The sincerity in his face made the whole thing seem even more ridiculous, and the sight of him rushing along like this would make Wendy and me giggle.

Even adult owls look funny when they run on the ground. Barn owls don't usually walk on the ground in the wild, as they are only there long enough to grab their prey, so their feet are specialized for holding on to a branch with the long talons curled below. On the ground, the claws push their long toes up in an awkward way, and the extra pads on their feet seem to interfere with walking easily on flat surfaces. They don't do a sophisticated little run like shorebirds or hop sensibly along the ground like sparrows. Barn owls tend to make a dramatic mess of it. This fits their personalities. Nothing is simple or straightforward. It's got to be complicated, messy, urgent, and goofy-looking, in spite of the fact that owls take themselves very seriously.

On the morning that Wesley followed me from the bedroom into the living room for the first time, I sat down on the carpet and he scrambled to the safety of my lap, overwhelmed by this big new room. But in minutes curiosity overtook him, and he hopped down and began to explore. Wendy ran to get the camera. When she zoomed in on his face, Wesley stopped, cocked his head heavily to one side and regarded the camera with open curiosity. Wendy snapped photo after photo. Wesley stared right into the lens with his head bobbing up and down, back and forth, turning one way and then the other, the perfect high-fashion owl model. Wendy and I laughed so hard, it's a wonder any of those photos were in focus.

Almost three months old, using adolescent, or toddler, perch.
Stacey O'Brien.

Despite his rapid growth, at five weeks Wesley still didn't look like an owl. White down covered his entire body, including his wing stubs and legs. He looked misshapen and sort of lumpy. Around his bottom and hips there was a funny-looking mass of poofy white down feathers that he would retain into adulthood. I called these his "bloomers," which is what they looked like. Owl bloomers have an important purpose in the wild. Their ultra softness efficiently traps the warm air against the body, providing extra warmth and fluff when the owl pulls his foot up to sleep during cold weather.

At six weeks, Wesley's large head still overwhelmed his little body. His feet looked entirely reptilian and scaly. And they were huge compared to the rest of him, with talons sharp as razors. At this stage in the wild, baby barn owls flap their growing wings hard and climb tree trunks by hooking their talons into the bark and powering with leg and wing muscles straight up the side. It's a great preflight exercise. Unfortunately, Wesley saw me as his personal tree. His beak and talons, meant to kill and rip flesh, really hurt. I started wearing thick jeans at all times, even to bed.

At my mother's house one day, she watched Wesley climbing all over my bare arms.

"Don't his tentacles hurt your skin, dear?" she asked.

"They're not tentacles, Mom," I informed her.

"Oh, I mean, don't his testicles hurt your skin?"

"I think you mean talons, Mom. They're called talons."

She shook her head and questioned how I'd ever find a husband with bird scratches all over me. This comment stirred unease about my upcoming date with Paul. While biologists are proud of their scars and love to trade what's-the-weirdest-thing-you've-been-bitten-by stories, Paul was a nonbiologist.

My arms were covered with long razor-thin scratches from Wesley's talons, but I also had some unique marks from other animals, including gouges from large owls at Caltech. But my coolest scar by far was on my right wrist from a three-foot-long benthic worm with a six-inch retractable jaw. (Likely the inspiration for the jaw that shot out from the monster's throat in the film *Alien*. It looks exactly the same.) *Benthic* is the term for the layer of sediment at the bottom of the ocean, where these worms live. Their jaws shoot up from the mud and snag prey as it scuttles above them. I was on a seagoing vessel studying the effects of the changing currents on benthic life (and also, concurrently, the zooplankton that floats on the ocean's surface). We had netted a benthic worm that suddenly clamped on to my wrist and wouldn't let go. I wanted to jump around the boat and scream, "Get it off! Get it off!" but with all those scientists around, that would have been so uncool. So I froze and choked out, "Can somebody help me detach this thing?" Finally we were able to pry the jaws open and throw it back into the sea, but it took a chunk of my flesh with it. Nonetheless, I was triumphant. I now had the best "bitten by" story of all.

Yet I wondered whether a benthic worm scar would be an asset on a date with a musician.

Wesley and I were developing a nighttime ritual. As I washed my face and brushed my teeth in the bathroom, Wesley stood on the counter. He watched while I turned on the faucet, ran the toothbrush under the water, and brought it to my mouth. It wasn't long before he participated in the routine by grabbing the toothbrush and then parading around the sink with it in his beak. I would take the handle and gently try to pull it out of his mouth, resulting in a slow tug-of-war with me saying, "Wesley, give it back. It's mine, Wesley, now, give it back," to which Wesley responded by pushing his heels forward and pulling with all his might, screeching all the while until I would have to give in. "Okay, you win" and I'd carefully relinquish my grip. Finally, I figured out that all I had to do was take one of "his" and pretend to use it, let him grab it, then use a new one in peace. The moment he won this pitched battle, he'd lose interest in the toothbrush and throw it over the side of the counter, watching it fall with the kind of fascination babies in high chairs display after throwing their sippy cups.

One night Wesley poked his head under the running water. The feel of it surprised him and he jerked back, shook his head, and stared up at me as if asking, "What just happened?" Then he tried it again. Wesley had discovered water. He loved it. In fact, he became obsessed with water, trying to jump into the sink whenever I turned on the faucet. So I found an old dog dish, filled it up for him, and set it on the counter. From then on, Wesley imitated my bedtime routine using his own "sink," swishing his face in the water while I was washing mine, and drinking while I brushed my teeth. This was surprising behavior, because it was well known that owls did not go near water, let alone wash their faces and drink. At that time, as far as we knew, no naturalist had ever reported observing an owl showing any interest in water, ever. Owls get all the fluids they need from eating mice. Wesley had not read the literature, however, and his doggie dish was a fixture in our lives from then on. The group at Caltech wanted to see this behavior for themselves, so I brought his doggie dish to work and they all watched him play, drink, and splash around in the water.

Wesley had begun to display a fascinating variety of facial expressions and body movements. They seemed to reveal a great complexity of thought and awareness. I talked to Wesley the same way Wendy talked to her infant daughter, Annie, using words and body language simply and consistently. I was reasonably certain that he would learn to understand me on some level, but had no idea to what degree. I'd use the same words for specific actions, like "Wesley, do you want some mice?" whenever he ate, or "Wesley, go to sleep," when I went to bed. And whenever he saw something he thought looked interesting, which was nearly everything, I would name it out loud. Then he'd race toward the object in his galumphing gait, wings out like an airplane, with me running behind saying, "Wait! Not for owls, not for owls!"

THE NIGHT OF the big date finally arrived. I stood at Paul's front porch with the little third wheel asleep in a box tucked under my scratched arms. Paul opened his front door with a smile, ushered us in, and then peered down at Wesley.

"That's an owl?"

Wesley was just waking up from his nap.

"Well, yes, it's a baby owl," I said.

Paul bent his head over the box for a second look.

"And are those . . ." he exclaimed with widening eyes, " . . . cut-up mice??"

"It's what he eats," I offered.

Paul turned pale and backed away.

"That is gross. That is really gross."

There was no going out to a restaurant. Paul ordered in a pizza and ate it perched tensely on the armchair away from the couch where I sat with Wesley. He nervously eyed the box as if Wesley or the cut-up mice were going to come flying out and attach themselves to his head. My visions of going down the aisle to "She Blinded Me with Science" faded quickly. I found myself wishing we were back at home, just Wesley and me, getting ready for bed.

Finally, Paul popped in a video and settled back in his chair. *Oh, good, he's finally relaxing*, I thought. Then I heard snoring. And it wasn't Wesley. Mom had been right. I hate admitting that.

I slipped out of Paul's house with Wesley, strapped his box into the passenger seat with the seat belt, and started the car. A blanket of sadness descended. I knew Paul would never call again. I'd been so sure he was "the one." What a bummer. Oh well, maybe there was something to be said for having Wesley as sort of a litmus test for guys. *Love me, love my owl*, I thought. I certainly couldn't be with a guy who didn't have some interest in animals. Paul seemed to have none. Wesley's eyes searched my face as if trying to decipher a work of art. Perhaps my little owlet had just prevented me from making a big mistake.

I pulled into the long circular driveway at Wendy's house. As I walked up the driveway, her flock of geese announced my arrival. The horses nickered, and a goat bleated. I could hear the chickens cackling to each other. The scent of hay was on the breeze.

It was a relief to be home with Wes.

We got ready for bed and had our nightly cuddle. Wesley lay on his tummy across my left arm, nestled against my stomach, head in my hand with his legs hanging over the side of my arm. With my right hand I rubbed him on the bridge of his nose, which I knew he loved, because he closed the disks of his face like a little taco over each eye, opening up the area over his nose for rubbing. With his "cuddle face," he fluffed the feathers immediately around his nose, so that most of his beak was hidden and all I could see was a little bit of the pink tip.

One of my tears hit the feathers on his back. "I'm all right, Wes." I told him, "I wouldn't be happy with a man who doesn't understand animals anyway."

Actually, Paul did call a few times over the next year or so. He'd always eventually ask, "Do you still have that owl?" He said "owl" as if he were asking, "Do you still have that human head hidden under the bed?" or something equally horrible. "Oh, yes!" I'd say and describe Wesley's antics to him. He'd sigh deeply and ask, "How long do they live in captivity?" I'd always answer, "Fifteen years or more." Then he'd say "Okay," and quickly end the conversation.

A few years later he got married.

5

Flying Lessons

Wesley pounced on everything from pillows to items on the floor. *Stacey O'Brien.*

ONE DAY WHEN he was nearing seven weeks old, Wesley drew his wings way up over his head, leaned forward, and had a good long stretch. It was then that I saw the beginnings of pinfeathers on his wings. *Pinfeather* is a catchall term for any new feather coming in. Although they're called pinfeathers, they are actually waxlike tubes made of keratin, the same stuff our hair and fingernails are made of, but they are alive, have blood and nerves, and are extremely sensitive. Inside each living tube, an adult feather is being formed. When the feather is ready to meet the world, the blood and nerves recede and a white waxy keratin sheath remains.

One of my favorite activities once Wes started growing pin-feathers was to groom him. As he lay on my left arm against my stomach, I'd gently pinch the white waxy part with my right hand, pulling it off carefully, to watch a perfect new feather emerge, fully formed. In essence, I was grooming him just as mother owls groom their babies and mated pairs groom each other. The adults can groom themselves, so they seem to be doing it to bond and show affection. Birds in general love to have their pinfeathers groomed, and we would do this for hours for the rest of his life, as even adult birds are constantly grow-ing new feathers, to replace old, worn-out ones. It was thrilling to watch the gradual emergence of Wesley's beautiful golden wings.

Spring was on the way. During this time of year I would occasionally go out to forests, malls, and countrysides to watch for wild owls. One moonlit night, I saw a barn owl emerge from his cubbyhole on the roof of a shopping mall. He flew several hundred feet up into the air before finally free-falling sideways. He did loop-de-loops and figure eights, then flew in circles ascending higher and higher until he was just a speck in the sky. He drew in his wings and dive-bombed straight toward the ground gathering great speed, pulling up only at the last second. I had never seen such a wild display of joy and prowess. I hadn't even known it was possible for an owl to fly like that, much less do so seemingly just for the thrill of it.

This owl was an adolescent, meaning he was less than three years old and, thus, still without a mate. Although most literature says that barn owls mature sexually at one year old, I disagree. Wesley did not reach maturity until he was three and a half years old. It was my understanding that we had made the same observation at Caltech. I had developed an intuition about the age of barn owls, based on my work with them at the lab, and found that the younger, unattached ones seemed to have slicker-looking feathers—they all looked brand-new. These owls spent their time on their own, not yet calling for a mate since they weren't sexually mature. Older, unattached owls have an air of depression about them and a scruffier mantle of feathers, compared to the youngsters. Maybe this one was getting ready to impress the ladies. Or maybe it just felt fantastic to fly with such abandon. Watching him took my breath away. His beauty and power humbled me and also made me a bit sad for Wesley, who would never be able to experience such flying.

By two months, Wesley had almost completely lost his white baby down feathers, which I carefully saved in a little box, because they were as precious to me as a baby's first curls. His adult down, insulating him under his visible feathers, would be a deep gray, not white. As he grew more and more long flight feathers, he'd start trying to fly soon.

When I adopted Wesley, I was also keeping zebra finches as pets and had had seventeen birds in twelve cages in my bedroom. I wasn't sure how well this was going to work once Wesley could really fly, since barn owls will eat small birds (up to about 3 percent of their diet) to supplement their normal fare of mice. For the first few months, I put curtains on their cages and kept them up high where Wesley couldn't get at them. I was optimistic that, since he was raised with them, he might leave them alone, but just in case, I started telling people that I might need new homes for my birds soon.

Wesley's first flying attempts involved stretching his wings and flapping them hard. This didn't do much to get him airborne, but it did stir up a wind that sent everything that wasn't weighted down in the bedroom flying through the air. Next he tried combining the wing flapping with short hops, which made him career into everything in his path.

The flapping-hopping-crashing stage went on for quite some time until one day he flapped and jumped a little higher into the air. This time, he stayed airborne. Wesley was flying! But he had no control whatsoever. He tried to steer himself, but the power of his wings overwhelmed his technical abilities. When he aimed for something to land on, such as the dining room table, his approach was like that of a jet trying to land on a short runway. He touched down on his rear, slid to the other end at top speed, and dropped off onto the floor in a tangle of wings, feathers, and feet.

I laughed. I couldn't help it. Then I ran over to console him, but he turned his head away and refused to look at me.

"Wesley, what's wrong?"

I checked him over thoroughly, but nothing was broken or damaged. I turned his head toward me and looked into his eyes.

"Wes?"

With a force I hadn't felt in him before, he jerked his head sideways and stared at the wall. "Are you okay?"

Wesley pushed me away with his wings and hissed under his breath. I let go, and he stood away from me and stared at the wall again. No matter how much I tried to comfort, cajole, or beg him, Wesley refused to look at me. With a chill, I thought he looked an awful lot like that owl that got lost in the ventilation system and willed himself to die. But I convinced myself this had to be different because all owls must go through the experience of learning to fly. I decided to leave him alone, and after he slowly groomed his feathers—with his back to me the whole time—he reemerged from the far end of the dining room table in as dignified a manner as he could muster. Then he tried again. He flapped and hopped until he was up in the air and, frantically looking around, eyed the dining room table again and headed its way. This time he stuck his feet out in front of him and held them open like hands trying to grab solid ground. But it didn't help. He hit the table, slid on his rear all the way across, and crashed on the floor again.

Again I dissolved in laughter and again Wesley stared stonily at the wall. I stopped laughing abruptly when I realized that Wesley was embarrassed. Learning to fly is physically and emotionally very difficult, and human owl mothers should not laugh at their babies. From then on I tried my hardest to keep a straight face.

Most pet owners know that animals can read emotions such as anger, approval, affection, and acceptance. But it had never occurred to me that perhaps an animal could feel ridiculed. From that point forward, no one in Wendy's house was allowed to laugh at Wesley, at least not in front of him, while he was learning to fly. Sometimes we had to run into the bathroom, shut the door, and burst out laughing.

Scientists are generally afraid to assert that animals feel such emotions as embarrassment, mainly because it's hard to prove through experiments and accepted scientific methods. More and more scientists, though, are beginning to believe that animals do have emotions and that their feelings may be more intense and unfiltered than our own. Emotion arises from the old brain, the limbic system, which birds and reptiles as well as dogs, humans, and other mammals share. Humans have additional brain structures and symbolic language to process our feelings and a complex array of psychological defense mechanisms that allay or soften the impact of our emotions. We repress, deny, subjugate, dissociate, and use all kinds of conscious and unconscious machinations to separate ourselves from our feelings, but animals have no such recourse, so their emotions likely are raw and strong. In fact, this may be one of the reasons we find them so attractive: they wear their hearts on their sleeves, so to speak. We understand all this intuitively because we can recognize emotions when we see them, as we share them with the other animals of our world. Since our own brains are of the same pattern as the brains of other animals, emotions are more likely to be universal to all creatures possessing a brain than they are to be unique to humans. People seem to deny the existence of animal emotions so that they can continue to justify inhumane treatment and exploitation and avoid the fact that our actions have a deep emotional impact on our fellow beings.

The evidence that all species of animals with a brain have emotions is overwhelming. I've observed that all intelligent animals have emotions, including reptiles, whose brains are less complex than those of mammals. People who work with reptiles are well aware of the risk of depression in captive snakes and lizards of all kinds. Turtles and tortoises are especially prone to it. If a snake gets depressed, his life is immediately in danger, as he will stop eating. I once rescued a snake that had to be tube fed for a year before he began to eat on his own again, after having an owner who did not provide proper stimulation for him. Snakes will also stop eating if they have a traumatic event with a mouse. Reptiles are cold blooded, meaning that they cannot control their own body temperature and are dependent upon their environment to provide a heat source. If they can't raise their temperature, their metabolism becomes so sluggish that they cannot defend themselves against even a mouse. Careless snake owners have been known to toss a mouse in with the snake and not supervise. If the snake is cold, the mouse can eat the snake alive and the snake can't respond. If the snake survives such an episode, it will have such a fear of mice that it will no longer eat. It can take up to a year of tube feeding before the snake gains the courage to face another mouse. If an animal of such low intelligence is this emotional, how much more does a highly intelligent animal feel? Even a reptile needs an "enriched" environment—and it's vital for more intelligent creatures. Animal keepers try to enrich the captive animal's life—make it more interesting—to prevent disorders like obsession/compulsion (incessant pacing in a cage is a good example) and depression. For instance, caretakers hide food all over the enclosure rather than simply putting it in bowls, so the animal has the fun and stimulation of hunting for it.

Lack of stimulation affects brain growth. The less enrichment in a rat's cage, for instance, the less the brain will develop. The difference between a rat with a wide variety of toys and one with no toys can be seen with the naked eye during an animal autopsy (a necropsy). The rat with the toys will have a brain just packed with ridges and wrinkles, which indicate more neural connections, while the rat with no toys will have a relatively smooth cortex because he lacked the stimulation needed to develop neuronal (nerve cell) connections in the brain. Of course, boredom affects the emotions of a captive animal, too, and depression is a serious problem.

The more intelligent an animal is, the more likely he is to have complex emotions. According to scientists who have worked with them extensively, parrots, as well as many primates, depending on the species, are about as intelligent and emotionally mature as a two- to five-year-old human. Pet owners and bird enthusiasts will tell you that a parrot can be so devastated by the loss of his owner that he can die of depression, much like the owl who wills himself to die after the loss of his mate.

I was worried enough about Wesley's extreme emotional response to our laughing at him that I banned laughing at him in all situations. This was a difficult rule to follow, as so many of his antics were impossibly cute and, yes, funny. The way he gyrated his head when he was interested in something was comedic. His head would saccade horizontally with his face in the same position, unmoving, like in an Egyptian mural; then he would twist his head all the way upside down to get a better fix on the object; and finally he would pull his head back then throw it forward at the object like a cartoon animal doing a double take. How could we not laugh when he did this? It was so cute! So we had to either stifle it or run into another room and burst out laughing in there. He was taking himself quite seriously, so we had to take him seriously, too.

After about a week of disastrous crash landings, Wesley finally figured out how to back-wing, a delicate maneuver that involves using the bottom part of the wing to push some of the air forward, which allows owls to hover. Now when he approached a desired landing spot, he hovered hesitantly over it like a helicopter for a moment and then slowly sank to a standing position. This feat was accompanied with much praise from me and great excitement within the household. When Wesley finally learned the last-minute-braking-and-landing method, Wendy and I erupted in cheers and Wesley joined in with a loud exclamation of excitement that sounded like, "Deedle Deedle DEEP DEEP DEEP DEEP deedle deedle," turning toward us with bright eyes as if accepting the praise, and flapping his wings.

I think that Wesley took longer than most baby owls to make most of the normal developmental steps. Perhaps this is because he had no owl parents to teach him and had to figure everything out for himself. The nerve damage in his right wing may have inhibited his ability to strengthen it enough to control his flying. Even after he had become quite adept at flying, he had problems with endurance in that wing and it would droop after a few minutes. In the wild he could not have hunted long enough to provide for a nest full of babies, and may not have been able to fend for himself. But he obviously enjoyed flying recreationally.

One weekend Wesley slept most of the day in my room, and as it grew dark I thought I'd better go check on him. I opened my bedroom door, and while I'm sure he meant to land on my head, he lost control, flew right into my face, and got tangled in my long blond hair. He panicked and scratched my face pretty badly. Once we untangled ourselves, I acted as if nothing had happened to avoid the big "I'm-so-embarrassed" scene that followed his flying mishaps. For weeks afterward, though, concerned people would ask me if I had an abusive boyfriend. I was so used to being scratched by animals, especially owls, that I never sought treatment, but just kept my tetanus shot up to date, as all biologists do. (Rehab centers recommend that their workers get the rabies shot, but few people actually do, myself included, as rabies is rare.)

When working with birds of prey, the slightest lapse in focus can be disastrous. Shortly after my collision with Wesley, I was at work getting ready to clean out the quarantine quarters of a playful great horned owl who loved to pounce on anything that moved. We had taken him in from another center, so we had to quarantine him to make sure he didn't have any owl diseases that could spread to the regular population. Great horned owls

Wesley has just learned to fly and is still unsure of his landings. *Stacey O'Brien.*

are huge. They are much larger, heavier (around five pounds), and more powerful than barn owls, who usually weigh about one pound, and can have a wingspan of four to five feet. This great horned owl was housed temporarily in a large wooden box–like structure with a perch and a sliding metal floor for cleaning. I carelessly reached my hand in under the door to slide the bottom tray out, and he pounced, burying his talons deep into the bones of my hand. I heard a long, high-pitched scream echo through the owl barn and realized it was my own. Although I was in extreme pain, I had to scoot my wounded hand along the tray, with the full weight of the great horned owl still on it, and move it close enough to the cage door to reach in with my other hand to detach him. Somehow I pulled each of his individual claws out of each of my individual bones. The owl still wanted to play and punched the closed cage door with his talons one time for good measure, making a clacking sound with his beak. Then he pounced along the edge of the tray, hoping for another shot at my hand.

This time, since I had more than a scratch, I was sent to a doctor right away through workers' compensation, which as the name implies, only deals with work-related injuries usually caused by the physical demands of manual labor. The doctor's office was filled with men in flannel shirts and work boots who had injuries like broken bones and ruptured disks. I filled out the form, handed it to the receptionist, and sat down to wait my turn.

Soon a grim-faced nurse appeared and asked me to come into her office to clarify some of my answers. Looking at the form, she said, "Now tell me again how were you injured."

I answered, "I was attacked by an owl."

She put her pen down and stared at me. "You can't put that here," she said.

"But, it's the truth . . ."

She stood up and excused herself. A few minutes later, she came back with a supervisor.

"Now, young lady, we seem to have a problem. This is not a work-related injury."

I wrangled with him for almost an hour and finally exclaimed, "Look, it's Caltech research. Just call my boss."

He drew himself up to his full height and said, "We just may have to do that."

He left the office and soon afterward the nurse reappeared and sheepishly gave me a tetanus shot and antibiotics and sent me on my way.

I had been lucky that there was a door between me and the great horned owl. Owls generally go for the eyes with their talons when they are truly attacking, and this owl might have done that. Although we biologists were encouraged to wear goggles, we usually didn't, since they were annoying and cumbersome. One guy who had been attacked by a barn owl thankfully was wearing glasses but still ended up with four neat puncture wounds around each eye. I don't think scientists know why, but all predators and most prey know to go after the eyes of another animal, and they seem to know where the eyes are, even on species very different from themselves. Spitting cobras are exquisitely accurate, perfect shots with the poison that they aim at other animals' eyes. Many animals—some moths, caterpillars, and fish—have fake eyes on less important parts of their bodies, usually their tails, so that attackers might be misled to missing their real eyes. It makes sense that those creatures who attack the eyes of an enemy will be more likely to survive and carry on their genes, and that most evolved to go after the eyes, which is the fastest way to disable prey or predators.

When I approach a wild animal, I keep my eyes averted and never look directly at it because the first thing it does is look at my eyes to judge my intent. But gazing into the eyes of tame animals is a form of showing affection. A hamster will sit in its owner's hand and look into his eyes as it washes its face: it will lie on a person's hand, nose to nose and eye to eye, content to just commune in this way. Even many types of lizards I've raised will look at each other—and would look at me—right in the eyes. It seems to be universally understood among all sentient animals, even reptiles, that the eyes are the windows to the mind of the being within.

After Wesley accidentally scratched me up in the flying incident, he and I moved into a bedroom at the far end of the house, to keep things a little safer for Wendy's family. It was perfect because it was large enough for him to fly more freely and had two doors between Wes and the rest of the house—the bedroom door that opened into a laundry and bathroom area, and another door that led to the rest of the house. Now, if Wesley were to escape from the bedroom, he would not be able to get into the main house right away.

Wesley perfected his flying in that room. He would jump off the bed, fly full speed across the room, and pounce on the couch with all his might. Then he'd grip the couch with his talons and flap his wings vigorously as if trying to lift it up. Sometimes he'd pop back up into the air, do a sort of flip, and attack the cushions. Sometimes he missed his target and ended up flying into the pictures on the wall, where he grabbed the frame and flapped his wings maniacally as the picture thumped against the wall. This was good practice and strengthened his wings, because he had to make a fast turn in order to fly away from the picture frame without falling. Soon he was good enough to zoom around the room and take the corners, just like a race car driver, banking up at the turns and making them by a hair's breadth.

At this age, Wesley had the power of a hunter but the awkward naïveté of a beginner. The first year of flying is perhaps the most vulnerable time in a wild owl's life. He doesn't have experience in the world, and he can't discern what's dangerous. This put me in a constant state of anxiety that Wesley would hurt himself. Nature helps the fledgling birds of prey, however, whose wing and tail feathers are actually longer during their first year than they will be for the rest of their lives. The longer flight feathers give them added stability but sacrifice maneuverability. It's exactly the way air force pilots are taught to fly—first in planes with large wings that are steady but not built for delicate maneuvers. As the pilot gains experience, he graduates to planes with shorter and shorter wings until he can fly the extremely agile fighter jets that can dogfight and turn on a dime in combat. Amazingly, nature gives this same advantage to beginner owl pilots. I have been unable to find a formal study comparing the lengths of barn owls' tails and wings at different ages, but after my experience with barn owls, I can see it intuitively. By looking at lengths of the tail and wing feathers in comparison with their bodies, I've always been able to tell whether a barn owl is about a year old or more mature. Formal studies done with hawks have proved this point.

One of the most difficult moves for Wesley to master was hovering. Barn owls don't hover for very long because it's exhausting, but they do need to be able to hover when they are spotting prey in an area where there's nowhere to perch up high and watch for it. And they need to hover near the nest to offer mice to their babies. The best hoverers in the world of birds of prey, I think, are northern harrier hawks. It's magical to watch them hover almost perfectly still above a meadow as they watch and home in on prey. They seem to have incredible endurance, and their technique is elegant, effortless. To hover, the bird has to beat his wings more or less horizontally, with the more and the less in perfect balance. Part of the wing has to keep the bird vertical, which involves up-and-down movement, and part has to keep him from going forward, which is a horizontal movement forward and back. It's a lot like humans treading water: part of the arm motions we use push the water down so that we stay above it (the up-and-down movement), and part of the arm movements are horizontal to keep us from moving forward.

Because hovering is complex, it takes an owl a while to figure it out. As Wesley tried it, he would shoot forward or fall down too fast as he tried to balance all these complexities in his mind and teach his muscles to respond. The more he worked on his flying, the more amazing it seemed that any young bird could master it so quickly.

I've had the privilege of "flying" a military flight simulator and found that I did quite well with the SR-71, which is huge, ungainly, and built for long, stable flights that are so high in altitude that they are almost in orbit. The SR-71 was our best spy plane during the Cold War. I probably couldn't have crashed it if I had tried. But when I switched to an agile fighter jet with short wings, I could barely get "airborne" without spinning out and slamming into the ground. I never did get it stabilized, which brought home the difference between long and short wings.

Although he made a lot of noise crashing into objects, Wesley's actual flight was completely silent. Owls are the only birds that fly without making any sound. Unlike other birds, they have extremely soft flight feathers with serrated edges that mute the sound of their wings so that no noise interferes with hearing their prey. This silent flight also enables them to sneak up on their prey. Wesley's feathers were so soft that I could only feel them properly by touching them to my lips—my fingers just weren't sensitive enough. The top of the flight feather is also velvety so that even the surface of the wing creates no sound. Secondary feathers are extremely poofy, almost like very large down feathers. I could take any one of Wesley's feathers and swish it through the air and it would not make the slightest sound.

Few things are as startling as an owl making a surprise landing on the top of your head. Because his flight was so quiet, the only way I could anticipate this was to listen for Wesley's takeoff. Usually he relaxed on one foot and then if he decided to fly, he'd hit the other foot on whatever surface he stood before taking off. After many episodes of tangled hair and talons, I learned to listen for that foot-slapping sound. I'd brace for impact and stand still because if he were going to land on my head, I preferred it to be a good landing, not a big mess. Wesley was quite gentle if he didn't lose his footing, because he knew it was me he was landing on.

Other aspects of owls' feathering are unique to them. Most birds have a white powdery substance all over their bodies, particularly pronounced in parrots and their relatives. Some birds (herons and pigeons) have a special type of feather that never molts, called a "powder down feather," which produces a powder that they spread all over while preening, but even those that don't have a special feather produce a powder that's quite obvious. Barn owls have none. By contrast, Wendy's cockatoo, Omar, was so powdery that every time he shook his feathers, a cloud appeared, and when we groomed him, our fingers would get a layer of white dust on them. When I groomed Wesley, however, my fingers stayed completely clean.

Wesley grooming himself. *Stacey O'Brien.*

Almost all other birds also produce oil in a gland (the uropygial gland) located at the base of their tails, which they preen over their feathers, conditioning them to make them last longer. Wesley would religiously pinch this vestigial oil gland with his beak out of some old instinct, even though these glands don't produce oil in barn owls. Then he would act as if he were spreading the nonexistent oil all over his feathers.

Since barn owls can't oil their feathers, they are at a terrible disadvantage because they're not waterproofed like other birds. They get soaking wet quite easily and can become so weighed down by their soggy feathers that they can't fly. If an owl isn't able to dry off quickly, it can shiver with cold and die. Perhaps this is why they evolved to live in hollow trees and others' burrows. Wet down is a miserable, heavy substance, as anyone who has had to camp with a wet down sleeping bag can attest.

At about a year old, Wesley plays in the sink. *Stacey O'Brien.*

WESLEY WAS BECOMING more and more active at night. He started pulling his perch closer to the bed, tugging at it laboriously and patiently, then climbing my bedcovers in an attempt to get me to play. And of course it worked, since I had to get up anyway to move the perch back. I'd release him from his leash and run my feet under the covers, making sure there were many layers of blankets between my feet and him, and he'd take chase, pouncing on them. During the day if I took a nap, I'd let him stay off his perch since he was usually sleepy and would join my nap time. He would find an elevated place in the room to sleep—a pile of books on the desk or the top of one of the finch cages—but sometimes he'd rest on my shoulder or the side of my head. One time, though, I awoke to find him asleep on the side of my head with his foot positioned so that one long talon curved down into my ear canal, almost touching my eardrum. I didn't dare move my head, but I reached up and, with my finger, slowly pulled the talon out of my ear. Then and there I decided that Wesley needed his own pillow where he had to sleep when I was napping. He took to the pillow like he'd had it forever, and from then on he knew that it was his place on my bed.

Wesley had been expressive from the day he opened his eyes, but now that he was literally "leaving the nest" at three months, he was becoming increasingly opinionated and emotionally invested in everything that went on around him. He also began to withdraw from everyone in the household except me, as an owl would do in the wild when he separated from his parents and began to seek a mate. Wendy could no longer just walk up to Wesley and pet him casually: he would back away and hiss or even threaten her with lunges and hisses. "I guess my days of snuggling little Wesley are over," Wendy said one

Sitting on top of a covered finch cage as a young adult.
Stacey O'Brien.

afternoon. "Do you think he'll let me touch him?" I said she should at least try, and I held on to Wesley, who was already tethered, as she carefully and slowly approached him, speaking softly in a reassuring tone. "There, little guy. So handsome. Little Wesley, you're okay . . ." and Wesley held still and allowed her to pet him. For the rest of his life he would allow her to pet him if she approached him in this careful, gentle way, as long as I was holding on to him so that he wouldn't attack. Perhaps it reassured him that he was under my wing in a sense.

Wesley quickly sorted out who he would allow to do this and who he would absolutely not tolerate—men, dogs, and people he didn't know as a baby (with a few exceptions). Poor Courtney the dog couldn't figure out why her little friend didn't like her anymore, but she respected Wesley's threats and stopped trying to come into the room. Wesley always seemed to prefer my mother and sister, who resemble me in voice and looks, and he would sometimes decide that he liked someone for reasons that I couldn't fathom, other than the fact that they had a deep affinity for animals, were very patient and soft in their movements and voice, and they just loved him dearly. One woman who became his babysitter years later would sit by the door and read books to him. He would always have a special trust for Wendy, and later, for my dear friend Cáit Reed, who also sounds and looks a little bit like me, and also has a deep understanding of animals. As long as people stayed next to the door, he was fine. He might threaten them, but he didn't go after them. If he wasn't tethered and could go to a high place where he felt safe, he would allow people to come into the room if I was with them, without attacking. But he would not let them come close to him and would fly away as soon as they started toward him, landing on the other side of the room. I never let anyone test his limits beyond that. The days of Wesley the trusting baby were over. He was now a young teen.

Yet, like many adolescents, Wesley still had embarrassing moments as he perfected his flying. His landings weren't so hot, and he still slid across my desk and whacked into the wall fairly regularly, but with each day of practice, the humiliation of learning how to fly soon gave way to the sheer delights of knowing how to fly. And Wesley would come to express his joy in flying in many different ways.

Attack Kitten on Wings

Wesley at about a year old, "killing" a film canister. *Stacey O'Brien.*

MY BEDROOM AT Wendy's was dimly lit, but otherwise looked pretty much like any other bedroom full of stuffed animals. One particular day, however, one of the seemingly stuffed animals had a menacing glint in its eyes. It was Wesley and he was crouching motionless, his attention fixed on a small object in front of him. Suddenly, his head started gyrating wildly from side to side, round and round and upside down. In an instant, he shot into the air and pounced on his newfound prey—a hair scrunchee—with all his wild might. He pounced again and again, flying into the air and slamming back onto it with clenched talons, killing it over and over again.

Finally satisfied that the hair scrunchee was thoroughly dead, he turned to the next victim, an unfortunate empty film canister. Wesley flew around the room to gain speed and hit the canister at full power. *Wham.* Wesley pounced again. *Whack.* He held down the canister with his talons and gave it a bite that would have severed its spine, if it had had one. Then he looked at me for the first time and let out his new victory cry—"Deedle deedle deedle DEET DEET DEET DEET deedle deedle dee!" I murmured approval and he took a victory lap. Soon he would be ready for his first live mouse.

OWLS HAVE NO pack, flock, or herd, so they are absolute loners except for their mate and offspring. Because of this inborn inclination, Wesley was strengthening his bond with me and had started to consider every other life form an enemy combatant. If anyone else even looked into the room, he had recently begun to exhibit a strange display, which I called his "owl no-nos," because it looked as if he were shaking his head no.

My first experience with owl no-nos occurred in the wild owl section at Caltech, when I came upon a fledgling barn owl standing on the floor of an aviary, hunched over as if injured, rubbing his beak on the ground. I immediately went to his rescue, but he did not respond. If anything, the beak rubbing became even more pronounced. Maybe he was choking. I got down on my hands and knees and put my face close to his to figure out what was wrong. I stayed in that position for quite a while and he never changed his behavior or acknowledged my presence. I thought he must really be in trouble and ran up to Dr. Penfield's office, gasping that we had an owl down.

"Stay calm," he said, "and describe the exact behavior to me."

"Oh, Dr. Penfield, he's all hunched over and his beak is almost touching the floor and he's rocking his head back and forth exactly as if he's saying 'No, no, no.' I went in and checked—"

"You went in?"

"Yes and I got down on the floor with him to see what's wrong and put my face right up to his beak and couldn't see anything—"

"You what?" He leapt up from his chair. "That's the last threat before ripping your face off, didn't you know that? Stacey, you are lucky to have your eyes. That owl was trying to tell you that he was going to try to kill you."

I suddenly felt weak all over. What a close call.

"Well, that seems like a stupid threat display," I mumbled, embarrassed at my ignorance. "The owl can't even see the object of his concern, so how could he be threatening it? And who on earth could possibly interpret anything that ridiculous-looking to be a threat?"

Dr. Penfield just blinked.

It's not clear if other animals recognize this behavior as a threat. To say the least, it's downright strange, which might be enough to scare off another animal; but perhaps it's understood only among owls. It seems to be specific to barn owls, a signature move of theirs, though all owls rotate their wings forward to make themselves look larger, puff their feathers up, and sway from side to side, swinging their heads back and forth as they look at their enemy, and hiss and clack their beaks. All this occurs before barn owls go into the no-nos, which is their final warning. They look a lot like the monster in the *Alien* movies when it would hunch itself and sway menacingly. Hissing seems to be a universally understood threat among all animals, and most animals puff themselves up in some manner when threatened. Think of swans and cats, who puff up and hiss. But the fact that barn owls don't make eye contact during the actual no-nos makes it seem a ridiculous threat display until you realize that their primary source of information comes from their ears. They are more likely to respond to the sound of an enemy tensing its muscles and shifting to pounce than they are to rely on visual cues like we would. It's very difficult for us to think the way a barn owl does, so his ways can seem pretty ridiculous—but they make sense to him, which is all that matters.

NOW THAT WESLEY was more mature and tended to clam up with anyone else, I was the only one who could watch his playful antics. I tried to describe his entertaining flying and daredevil maneuvers to my friends, but they didn't quite believe me. One friend, Kurt, begged me to find a way to hide him so that he could see Wesley play. I finally gave in to his pleas, figuring that if I were to carefully ensconce Kurt in my bed under lots of blankets and he were to keep perfectly still, just like a field biologist observing wild animals from a blind, then we could bring Wesley into the room and Kurt could watch him do his thing. Kurt agreed to my brilliant plan, saying, "If it works with me, then you can do this for anyone else who wants to see him play."

I put Wesley in the bathroom by himself while Kurt quietly sneaked into my bedroom. We piled blankets on top of him and fashioned a little peephole.

"Okay," I said, "whatever you do, don't move or make any

sounds. And remember, patience is the key! He will start to play and if you can't see something, don't move, just wait until he comes back into view, okay?"

A huge Norwegian guy from South Dakota who could carry a refrigerator down a flight of stairs all by himself, Kurt suddenly looked worried.

"What do I do if something happens?" he asked.

"What could happen?" I said impatiently. "I'll come back in about twenty minutes. If he isn't playing, we'll try again another time."

Once Kurt was comfortable I said, "Okay, now be completely silent. I'm going to go get him."

"Okay," he whispered.

I placed Wesley inside the door of my bedroom and closed it like I always did when I let him fly around and play by himself. I didn't stay in the room because I thought that if he noticed someone breathing under the covers, he would think it was me, hopefully forgetting that I had left the room. I wandered off into the house and hung out with Wendy and Annie for a while, losing track of the time, so it was more like forty-five minutes later when I decided to check on how things were going. I stood at the door hoping to hear the sounds of Wes playing. I didn't hear a thing. It was dead silent.

I opened the door to see Wesley standing on the bed with his face less than an inch from the peephole, his body crouched, wings flung all the way out and rotated forward in the classic owl threat posture. He was rocking from side to side doing his no-nos, and occasionally lunging with a loud hissing snap of his beak.

From under the covers I heard a tiny voice, "Help me . . ."

"What are you doing, Wesley?" I said, and picked him up

and leashed him to his perch. Kurt then threw the covers back. He was sweating and pale, almost green.

"Where were you?" he demanded.

"Uh . . . well, I just thought you'd be okay. When did he discover you?"

"When? When? The second you shut the door, that's when! He came straight at me and has been threatening my eye for the last two hours!"

"Kurt," I said, "It's only been about forty-five minutes."

"Forty-five minutes? You try having an owl snapping at your eye for forty-five minutes sometime!"

I held back my laughter. "I'm sorry, Kurt. I was sure it would work out. At least we tried."

I should have known that Wesley would figure out that the person under my blankets wasn't me. I had underestimated his intelligence by a long shot. After all, if owls can hear a mouse's heartbeat, Wesley could have heard the heart of a big guy like Kurt and recognized it as different from my own.

Kurt made a long circle around Wesley's perch on his way back to the door and scooted out as Wes flung himself forward at him with one last snap of his beak.

BECAUSE WESLEY'S FLYING had improved, he needed an adult perch. I bought a parrot perch and modified it so that it was about four and a half feet tall, with a wooden dowel across the top for Wesley to stand on and a round platform below about three feet in diameter. In the center of the dowel I added a leather attachment for his leash, which turned freely so that nothing would get tangled, and I used strips of leather and chicken wire to block the area below the dowel so that Wesley

About eight years old on his adult perch, Wesley enjoys a mouse dinner, preparing to swallow it headfirst. *Stacey O'Brien.*

wouldn't wind his leash around it and get pinned. I covered the platform with towels and taped them into place. The nesting box was history now. Wesley developed an entirely new play routine on his perch.

Late one night, 3:00 a.m. to be exact, I should have been sleeping, but Wesley was up and it was far more entertaining to watch him play. He knew the exact limits of his leash and flew in a circle just within that perimeter, playing "helicopter," then flinging himself at the perch, power punching it by throwing his legs forward in an arc so that all the muscles of his chest, legs, and feet were fully engaged when he hit the dowel. *Pow!* What a move! I had added that resounding smack to my subconscious list of sounds that meant all was well in our world. Then he jumped off the dowel and onto the platform, with a hollow *pom!*

Then quite deliberately he dived over the side of the platform and hung there dangling on his jess like a huge golden bat. Enjoying the view upside down, he looked over the room from that perspective. The first time he did this I had tried to help him right himself, but he was so irritated that I did not intervene again. Once he tired of this view, he reached up, grabbed the towels with his talons, and powered himself back up onto the platform using his abdominal muscles, like a gymnast pulling himself up onto a bar. He would do this for the rest of his life just for fun, it seemed. I'd walk into a room and there he'd be, hanging upside down, looking around calmly. I sometimes did lift him back up to the perch just to make sure he was okay, which he always was. The only time I've known of an owl doing this in the wild was from a curious picture of a spotted owl hanging from a branch with the same expression of enjoyment that Wesley had. The person who took the picture said the owl wasn't stuck and easily righted itself when it got tired of hanging that way. Ravens and crows do this also, seemingly for fun.

Television documentaries about animals all echo that they play in their youth because it's necessary for them to learn the skills they need to survive—building muscle, coordination, fighting, and hunting moves. But this conventional wisdom doesn't begin to explain the full picture. Some species never seem to play at all, yet they still survive in the wild. On Wendy's little farm, we had chickens, geese, cockatiels, and a cockatoo, among other animals, and we observed them closely at every stage of life. Annie's pet chicken Herkomer never appeared to play, although she was as tame as an animal could be. Other species of birds play all their lives, like birds of the parrot family—and owls—long after they've learned adult skills. Wendy once knew a woman who kept a screech owl that also played like a kitten with wings all of its life, well into old age.

I actually don't think play is related only to predation, as prey animals such as rats, mice, ground squirrels, and goats play as adults, and some predators such as owls, otters, cats, and dogs also play as adults. At a wildlife refuge where I volunteered, grown Bengal tigers played for hours in their pool. One of them would carry his favorite red ball everywhere, throwing it ahead and pouncing on it, tossing it into the water then jumping in after it, pushing it all over the surface with his nose. Tiring of that, he'd take it out of the pool and roll on his back, batting the ball into the air like a kitten, using all four paws to keep it in place. His enclosure mates would join in the fun, rolling around and growling like Harley-Davidsons revving their engines.

So what makes one type of animal playful whereas another isn't? Or are they all playful, but we are unable to recognize it when some species do play? Clearly it's not simply that animals play just to learn certain skills. There needs to be more scientific study on why animals play or don't play without the cultural bias that many western scientists seem to have against ascribing "fun" or "joy" to animals for fear of seeming to anthropomorphize them.

Wesley was as playful at age thirteen as he was when he was a year old. Of course, I could say that he learned a certain behavior in order to experience the infusion of endorphins that play released, but really, is this necessary? Maybe he just did it for fun. Where did our emotions come from if not from our animal ancestors? Many human emotions are similar to those of animals. We are them and they are us.

When off his leash and flying around the room, Wesley dive-bombed the pillows on my bed and popped straight back up, barely losing speed, then zoomed across the room to smack his talons full force into the couch, leaving punctures wherever he landed—an attack kitten. Then he headed straight for the wall and landed on the edge of a picture frame, holding on with all his might and flapping his wings so hard that the frame banged against the wall over and over again, leaving a dent. He finally lost his grip, rescued himself from falling with a quick flip sideways and a swoop to the floor. There he explored every nook and cranny, including suitcases, boxes, and bags. Galumphing with great intensity toward the dark closet—I had left the door open—he slipped inside.

For a long time I heard him rummaging around. Then nothing. Hmm, awfully quiet in there. I started to worry and finally crawled into the closet to investigate. There was Wesley holding perfectly still, doing a split, with one leg desperately clinging to the top of one dress, the other leg holding on to another. Wesley was stuck, and uncharacteristically quiet about it. I recalled that, in the wild, when animals think they're in trouble, they immediately freeze, hoping not to attract predators. He may have been anxious in that relatively new territory, but he allowed me to rescue him and return him to his perch.

That wouldn't be the last time Wesley decided to mountain climb the clothing in my closet. From his vantage point on the floor, looking up at the dark spaces between clothes, it probably felt familiar to his instincts, which had evolved so that he would expect to be inside a hollow tree, or be attracted to that vertical dark situation. He'd explore this "tree" by working his way to the top, bracing himself between my dresses as he ascended. His legs would inch apart a little more at each step until he could no longer move, then he'd hold on for dear life with his talons. Finding him in that helpless position, I'd give him a little nudge so he could finish climbing up.

Soon, every piece of fabric I owned bore little punctures, as if someone had patiently gone over each garment with an ice pick. I figured that being a biologist was explanation enough, so I just walked around in clothes with lots of holes in them. As a rule, scientists don't seem interested in fashion and don't tend to worry about what other people think, outside the machismo of the science culture. If I had taken a poll around Caltech, few would have known what Prada was, much less which shoes were worn in what season. If we shower and shave, we figure, what more do you want from us? One of my office mates owned ten identical pairs of gray pants, white shirts, and white socks, to go with one pair of Birkenstocks. For decades, he wore the same outfit to work and never needed to shop or make a decision about what to wear. His was a fairly typical and perfectly acceptable style among scientists.

Wesley left a mark on almost every item I owned, and I sacrificed more and more of my property to Wesley's whims—clothes, books, papers, blankets, and furniture. I just didn't care about those things and felt like the luckiest person in the world to have him in my life. Each talon puncture was evidence of my baby's brilliance and personality. I actually saved books that he'd ripped apart because his beak marks were on them. I was pretty besotted.

Wesley's play included me of course. Our early game of his chasing my feet under the blankets had evolved. At first he had just run back and forth on the bed chasing my toes, but as his flying improved, he would fly straight up into the air, flip, then do a power dive and thrust his legs forward for the attack. Then he would fly a lap around the room to gain speed and pounce with all his might on my well-blanketed feet.

Of course there were little mistakes. One time my foot slipped out from under the covers while I was napping, and he did his power dive into my bare foot with full talon crunch. I leapt straight up from my pillow shrieking in pain, scaring Wesley so much that he flew around the room in a panic until I calmed him down and was able to hold him. He started his "I'm so ashamed" routine of pushing me away, refusing to look at me, hunching up, and trying to face a wall. But I comforted him, crooning to him, "It's okay, it's okay." While I cradled him, my foot was bleeding, but that would have to wait. It wasn't his fault, I was the one who had let him play this way for so long.

Since owls don't flock, herd, or pack, they have no social setup for correcting each other's behavior. Therefore, Wesley had no way to interpret any act of aggression except as a threat on his life. For this reason, the number one rule in interacting with birds of prey is that you can never show them any aggression. You cannot try to discipline or correct them as you would a child or dog. They would not understand it. I could never raise my voice or do anything that might seem at all aggressive, even when trying to stop Wesley from doing something for his own protection. I could only gently remove him from whatever situation was putting him or me in danger. Eventually, he might learn that a certain behavior wasn't allowed, but not in the usual way. It took longer and required much more patience than the normal pet owner or parent is accustomed to.

Wesley had his own mind and did not obey me, but rather, lived in a kind of mutual cooperation with me. I did not actually try to train him in the traditional sense, but simply endeavored to protect him from himself. He learned from me as he would have learned from an owl parent—not by being corrected, I think, but by observing me and making his own decisions about how to behave. Certainly I was not interested in changing him, since I was learning from him about what it meant to be an owl. Taming an animal is not the same as training him. I am not an animal trainer.

Social animals in general are easier to teach because they seem to think the same way we do; they understand that a show of aggression is just a temporary correcting reaction. Wolves growl and snap at pack members, saying, in effect, "Back off," so that they avoid fighting and particularly avoid fighting to the death. They understand aggressive signals. Owls don't. Owls will think you are trying to kill them, and for the rest of their lives they will remember that you tried to kill them. So if you mess up even once and yell at a bird of prey or, God forbid, make a threatening gesture toward it, it will never be as tame with you again. The owl has no context for such frightening behavior. That's it. There are no second chances in the wild. That is the Way of the Owl.

One of the reasons I had moved in with Wendy was to help out with her baby, Annie, while her husband was on the road. It was a joy to be involved in helping to rear a child, since I did not have children of my own. Following Wendy's lead, we never said Annie's name in anger and we didn't say "no" unless it was a very serious matter. We just informed Annie, "Not for babies" when she reached for something unsafe, and reserved the word *no* for the most dangerous situations.

Knowing that I had to be as careful with Wesley as with Annie, I tried Wendy's strategy with him, too, and I avoided using the word *no* for minor issues. Whenever he got into something that wasn't safe, I said, "Not for owls" and removed the object gently from his grasp. Because he tended to lock on to anything that interested him, I'd often have to distract him to get him to let go. Trying to get him to let go of a mouse, for instance, would have been almost impossible if I hadn't distracted him, and he would have fought me for it. Thankfully, there were only a few occasions when I needed to take a mouse from him, for instance, when it wasn't fit for him to eat.

Annie came to understand the seriousness of "no" and would stop dead in her tracks if one of us said it. Wesley, however, only stopped long enough to consider whether or not he thought it was worthy advice. Although he knew exactly what it meant, he was still an owl—stubborn and wild.

Most people prefer to work with animals that have a social instinct, because they are more malleable. Owls' willfulness often causes them to be misunderstood. I was told that owls were "stupid," but much later learned that the person who told me this really had meant you can't train them to do your bidding. Well, those are two very different statements. Owls are highly intelligent. They just keep their own counsel and don't care to obey anyone else. Why should they? In the wild they are loners, although to their mates they are the sweetest, most devoted creatures on earth.

During Wesley's first year, I continued to follow him around as he played and explored, watching out for anything that could pose a threat. Danger lurked everywhere, and as his owl mother, I had the job of saving his life on a fairly continual basis. Besides "child-proofing" everything, I was extremely careful about what I allowed into the living space we shared. Drinking glasses and flower vases were forbidden because he might knock one over and cut himself. Besides the obvious poisons like drain cleaners and bleach, other dangerous liquids included all caffeinated beverages: the caffeine jolt we depend on in the morning can cause heart attacks in birds. I had no idea what might lurk in houseplants, so I just kept them out of the room. No plastic bags—Wesley might have put his head into one, gotten stuck, and suffocated. Eventually, almost all my framed pictures were taken down, as his flying attacks on the frames sometimes ended with his falling pretty hard to the ground with the picture. Anything on which he could land that would flip up in his face was also forbidden, such as a saucer or plate. Any important paper items were also eventually hidden in drawers so he wouldn't rip them to shreds.

Living in earthquake country meant I had to keep furniture bolted to the studs in the walls and had to make sure that nothing heavy that could fall on him was on a wall near Wesley's perch. Earthquakes usually start small, so even a minor quake

A few years old, playing in newspapers
next to his emergency carrier. *Stacey O'Brien.*

would set everyone into crisis mode. I kept an animal carrier under Wesley's perch and would race to put him in there at the first sign of a temblor. He seemed to understand what was going on and stayed in the carrier quite happily until I felt that the danger had passed.

Before I knew it, Wesley was celebrating his first birthday, February 10. Dr. Penfield and I figured that he was four days old when I first saw him on Valentine's Day, so I made the tenth his official birthday (or hatch day). I decided it was time to let him kill his own mouse in honor of all he'd learned and accomplished. After all, he'd been practicing his flying and pouncing moves, so why not try them out on the real thing? I did not take this lightly and stood by to intervene if Wesley started to hurt or scare the mouse without killing it instantly.

I got a small mouse and put it into the shower where it wouldn't hop out. Wesley would be able to take his time and chase it down. I imagined he'd pop into the air, do a fast swoop with the talon-crunch follow-up, then bite the neck just as he had bitten the film canister. No such luck. Wesley was terrified of the live mouse, cowering against the shower door and hissing under his breath. Gathering his last bit of courage, he finally faced it and gave it his "no-nos," standing in front of it and slowly moving his head from side to side. Not intimidated, the mouse simply washed its face. Dr. Penfield had said that an imprinted owl could never learn to hunt, so I should have expected this. The reasons for this belief are complicated. The most obvious is that an owl imprinted on a human doesn't have parents to teach him, and he was imitating me. But his threatening of the mouse was interesting. Perhaps he was confused, since he was used to seeing dead mice. Seeing one alive and animated may have frightened him so that he reverted to a threat gesture. But there may have been a lot more to it than that.

Humans and other mammals sense danger using a very busy part of the brain called the amygdala, which handles sensory input. It acts as a sort of Grand Central Station for interpreting what the senses perceive, using memory and emotions in the process. In birds, the amygdala has developed to become much larger in proportion to that of mammals, and may partially account for the extreme intelligence of birds as well as some of the differences in their thinking. Scientists used to think birds' brains were simpler than those of mammals; now we think they may be just as complex, but in a very different way.

Birds make the same kinds of neural connections as mammals, but certain parts of their brains have developed differently. In birds, the larger amygdala may have evolved to handle more complex processes than it does in mammals. In mammals, this structure is a primary player in the handling of emotion, but in birds it is used more to integrate information and less if at all for emotion and danger assessment.

The bottom line is that birds' brains have evolved a different intelligence from our own and we are just now starting to get a handle on how they use their brains. By comparing the brain functions and structures of birds and mammals, we can now see that their brains evolved in different physiological directions. Even with different structures taking on different functions, however, the two groups developed similar kinds of intelligence—so similar, in fact, that we can communicate and share emotions with each other.

Whatever was causing Wesley's strange response to the mouse, I decided to wait until next year to try it again.

7

Love to Eat Them Mousies

At about four years old with Stacey, Wesley is threatening the photographer. *Connie Fossa.*

SOME VERY IMPORTANT PEOPLE from England were scheduled to stay at Wendy's house during their visit to the United States. Our extended family was nervous and wanted things to go perfectly, so we spent an entire week cleaning the house, planning meals, and then cleaning some more.

On the evening the guests were to arrive, I brought home

two bags of live mice and set them in my bathroom so I could kill them on the floor and hide them in the freezer before the VIPs made their appearance. Like all full-grown barn owls, Wesley ate three to four large whole mice per day, and it was hard to keep up with his needs. However, before I could even begin, Wendy called me from the kitchen, "Hey, Stacey, could you help me cut up vegetables for salad?"

"Sure!" I replied, not wanting to leave her shorthanded. I rushed out of the bathroom, figuring I wouldn't be gone for more than a minute or so. More about that later.

As I chopped away on a bunch of carrots, I worried that if Wendy started telling them about Wesley, our visitors would want to meet him. If I leashed him to his perch, people could stand in the doorway and look at him, but once people saw how beautiful and expressive he was, they usually became insatiably curious, asking if they could watch him play, fly, and groom, which they could not do, as Wesley was particularly shy around strangers. I didn't blame people for asking, though. After all, it's a rare opportunity to get up close to an owl. I didn't mind showing him off, but I wasn't sure how they'd feel about his eating habits.

Wesley had two methods of eating mice. The first involved swallowing the dead mouse whole, headfirst, as a snake does. The pink tip of his beak peeked through his feathers daintily, but in reality his beak was huge, extending almost all the way back to his ears. One way I could easily wow people was to put my finger in Wesley's mouth so that it opened a little and they could see how it dominated his face and could easily take in a whole mouse. Wesley usually employed the whole-mouse method when he was very hungry or in a hurry, although he would first nibble around gently, without taking any bites, to arrange the mouse so that he could swallow it headfirst. Owls do not chew their food but either swallow a mouse whole or rip it into chunks, which they also swallow whole. But they can still breathe in spite of a mouthful of mouse because the glottis, the tube through which they breathe, is close to the front of the mouth near the base of the tongue. The esophagus is completely separate and far enough away from the glottis so that an owl can still breathe even with a mouse partway down his esophagus on the way to his stomach.

A very large mouse can take a long time to go down, in which case Wesley would produce a lot of saliva and sometimes even paw at his beak, unable to get it down or push it back up. That would really scare me and I would pull the mouse out by the tail. I learned to give him mice small enough to avoid this problem.

Hawks and eagles will pull the fur and feathers off prey before eating, but barn owls need that roughage. Fish-eating birds such as pelicans and seagulls have the same anatomical setup, with the esophagus along the back of the throat and the glottis up near the tongue and well out of the way of incoming food. Barn owls do not have a crop to store food as do seed- and fruit-eating birds and some raptors.

The other way Wesley ate was to pick out his favorite mouse parts, in this order: the head, heart, liver, and rack of ribs, which he plucked out through the mouse's neck hole after having eaten the head. If I gave him a large pile of mice he might go through and eat just the heads of every single mouse and leave the rest. A lot of the owls at Caltech and raptor rehab centers do this when there is an abundance of mice. Later, if they get hungry again, they'll go back for the most blood-filled parts, like the heart and liver. After all those good parts are gone, they'll eat the rest.

Wesley would toss whatever mouse parts he didn't want to eat off the ledge of his perch. Of course, this mess o' mice might land anywhere in my room, bed, or even on me. Wesley always hated eating the intestinal tract, and when he was a baby I'd disguise the guts by wrapping them in fur with some liver sticking out to fool him into swallowing them. He needed to eat the whole mouse in order to stay healthy, although I probably worried too much about it. Owl mothers don't try to trick the babies into eating everything. But when Wesley was feeding himself and being finicky, he'd extract the entire digestive tract with the skill of a surgeon and fling it as far away from himself as possible. Sometimes he would toss aside an entire mouse, which I'd pick up and return to his perch for him to eat later when he was hungry. Occasionally, I wouldn't see these food rejects and I'd end up stepping on mice—usually in bare feet—with guts popping out and exploding across the floor or over the top of my foot. I can't count the number of times I left the house with mouse gore on the bottom of my shoe.

Given this state of my owl's territory, I was nervous about our guests looking into my bedroom at Wesley and seeing mouse guts on the floor. Of course, I cleaned up regularly, laying towels below the perch and replacing them constantly. But sometimes I would miss something.

An enthusiastic eater, Wesley never tired of the same fare, day in and day out. When I brought him his daily serving of mice, he rejoiced every time and ate with gusto. If he was really hungry, he'd make intense begging sounds, pouncing on the mouse immediately, devouring it in an instant. If he was only mildly hungry, he'd take the time to sound off with a joyous exclamation, similar to his victory cry. Holding the mouse in his beak, he'd lift up his head, and celebrate loudly—"deDEE deDEE deDEE." Sometimes he became so enthused that he'd drool his acidic saliva all over the mouse, which would then slip out of his beak and onto the floor. Then he'd start crying and I'd have to pick the slimy mouse up off the floor and offer it back to him. He'd then repeat the whole process with his song of rejoicing, eventually getting around to swallowing his meal.

I have never seen any creature enjoy his food as much as Wesley did. He even developed a special sound while eating that I have never heard any other owl make. Holding a favorite piece of food in his beak, he'd emit a long, soft, secretive sound, a low staccato, rising slightly in pitch then remaining on one note for a long time—"dododoDoDoDoDoDODODO"—until it slid back down to the original note, "DODODoDoDoDoDododo," so soft at the end that only the two of us could hear it. He did it so rarely that I could never catch it on tape.

American barn owls' entire physiology is based on a mouse-only diet. Years before, at Caltech, when we had been using chopped rat patties, we had a tragic incident where an owl had choked to death on a rat skull. The rat head apparently got stuck in the esophagus for so long that the owl couldn't get it to go down or up and may eventually have choked on his saliva or died of the stress during the night before we found him. Normally, barn owls eat only what they've killed, and don't end up in this situation. American barn owls are unable to kill adult rats, which are too large and powerful for this fairly small owl to take on. Other bigger owls, that take much bigger prey like skunks, probably tear off pieces to eat rather than swallowing them whole. We realized that there were absolutely no shortcuts to feeding owls mice, food they would have killed for themselves in the wild, and from then on that's all we fed them.

People often asked me, "Couldn't you feed owls nuts or make them vegans?" No. Owls evolved to eat mice, just as cats evolved to eat meat and cannot tolerate a vegetarian diet. Dogs and cats evolved to eat meat plus the partially digested vegetable foods in the stomach of prey or scavenged carcasses; unlike owls, they need some vegetables as well for a balanced diet. It would be impossible for a true predator to live without meat. Wesley would not have even recognized anything other than meat as food. Putting a spear of broccoli on his perch would have been like putting a rock on my dinner plate. You cannot change the needs and physiology of a wild predator whose body evolved to eat a particular animal in a specific niche.

Predators play an important role in the balance of nature, keeping the larger numbers of prey animals in check. Because predators tend to eat the weakest of a species, they keep the remaining population strong. Without predators, herds become weak and disabled. In contrast, when humans hunt animals for trophies, they kill the strongest of the species, thereby weakening the herd. The famed Canadian naturalist Farley Mowat proved this in his brilliant study of wolves and caribou herds, which became the basis of the book and movie *Never Cry Wolf*. His study, in the 1940s, was so famous that it helped to bring an end to the inhumane practice of chasing down wolves by airplane and shooting them when they were too exhausted to continue. The Canadian government had hoped to justify this by saying that the wolves were decimating the herds, which they were not. Unbelievably, as I write this, the same practice has been made legal again in the U.S.A., and history is repeating itself in spite of everything we've come to understand about predators and prey. "Hunters" are shooting wolves from airplanes again—the very same wolves that biologists have been painstakingly restoring to their natural habitats over the last few decades, at great expense and sacrifice.

I do not condone cruelty or the wholesale slaughter of animals, but a wild predator's nutritional needs cannot be changed.

Some people feed their cats a vegetarian diet, but then let the cat outdoors (which they shouldn't, since cats are a leading cause of the decline in songbirds, after habitat loss and the continual poisoning of the environment), so that the cat hunts mice, rabbits, birds, and lizards to compensate for its dietary deficiency. If the "vegetarian" cat is kept indoors, he first goes blind, then dies of complications due to malnutrition. I never got used to having to kill mice and was just as horrified by it after many years as I was the first time. I had thought I would become desensitized to it, but it remained painful to do. Eventually, I found a pet store that would kill the mice for me before bagging them up, which helped somewhat.

I tried feeding Wesley chicken when mice were scarce, since at least chickens are related to the small birds that owls will sometimes eat. But he usually stared at it for a long time and then cried, begged, and pointedly stepped all over it until I took it away. I persisted because I needed some kind of mouse substitute to which I could resort if I just couldn't find any mice to feed him. Through trial and error I found that he would reluctantly eat chicken livers, gizzards, and kidneys cut into small pieces, with dark muscle meat mixed in. However, I could not feed Wesley chicken parts for more than a day or two before he'd become listless. For him it would always be an unnatural diet.

I could tell just by picking Wesley up that he had eaten a mouse, since he was so incredibly light that even the weight of one mouse made a noticeable difference. Also, his stomach pooched out in a perfectly round lump while the mouse was being digested. An hour or so later, when he was ready to cast a pellet, Wesley would lean to one side, looking more and more nauseated, then dramatically gag up the pellet, sometimes shaking his head to dislodge it from his throat. He seemed so miserable, but afterward, he'd give his head one last shake and smack his beak, seemingly glad to get that over with. For each mouse eaten, he produced one pellet. Rarely, he might eat two mice and produce a huge—2½ inches long by 1 inch wide—double pellet. As a baby, when he was eating six mice per day, he produced impossibly large pellets that must have contained, at times, up to three mice. I wish I'd kept them.

AFTER I FINISHED helping Wendy in the kitchen, I raced back to my bathroom to finish off the mice. But when I got there, the paper bag was open. My stomach tightened. No, it couldn't be. I counted the mice crawling in the bag, over and over. There should have been twelve, but two were missing.

I had to find them. And I couldn't tell anyone, not even Wendy, because this visit was too important and she was already anxious. Her husband would have been particularly upset to know that mice were loose. He barely tolerated Wesley, but was on the road so often that he didn't have to deal with Wesley's presence very much. I quickly killed the remaining mice, flicking them against the floor, and sneaked into the kitchen to hide them in the back of the freezer. As I headed back to the bathroom to search for the escapees, our guests arrived.

While introductions were being made they probably thought I was a nervous sort since I couldn't stop my eyes from darting around the room. It was such a large house that I knew that the longer I delayed, the less chance I had of finding the two mice.

Throughout dinner, the guests told us stories of their worldwide travels. My mind wandered, imagining mice appearing from just about anywhere: running across the table, scampering over our shoes, darting across the kitchen floor, dropping from the chandelier into our salad.

"Are you okay?" Wendy asked as I helped her serve the dessert course. "You're so quiet tonight."

"Oh, I'm just enjoying listening to all the stories," I replied.

After dessert, I politely excused myself from the table and hurried off to my end of the house. Almost immediately I heard one of the mice moving around deep inside the clothes dryer, which was located in the hallway between my room and the bathroom. Okay, I knew where he was. Now I just had to locate the other one. I was feeling quite hopeful that it would be close by.

I hunted for it most of the night, long after everyone else went to bed. I didn't want to sound like an intruder, so I tiptoed around the house with a flashlight, peering in every nook and cranny. I still couldn't find him.

Finally, before dawn, I quietly disassembled the dryer and extricated the first mouse, putting him in a box. Then I reassembled the dryer and slipped back into my room as the sun came up.

That morning we all gathered at the breakfast table, and everything seemed normal. After finishing their meal the English couple returned to their room and I started hunting once again for the second mouse. It didn't take long before I found a tiny, black poop in the hall outside the guest bedroom. Gulp. I had probably missed this telltale clue the night before. As I crept down the hall searching for another mouse dropping, the door to the guest bedroom opened.

The lady looked at me and then at the box in my hand.

"Excuse me," she said softly. "Have you lost a mouse?"

"Well, yes," I muttered in agony. "It's for my owl."

"I believe you'll find him in here. He seems to have made

himself quite at home." The mouse was sitting quite contentedly in the middle of her bed, washing his face. I scooped him into the box.

"You know," she said, "many people keep owls in England."

"Really?" I said, feeling profoundly relieved. "So you're not upset?"

Her husband chuckled, "Not at all."

She winked at me and closed the door.

I never fed Wesley mice that had been on the loose because of the chance that they may have gotten into something that could make him sick. Besides, it just seemed like these two had earned their freedom as a matter of principle. I took the mice on a short hike to a field near a stream and let them go.

8

Understanding Each Other: Sound and Body Language

IN SPRING OF 1986 I became very ill and had to be rushed into emergency surgery. While I was in the hospital, Wendy took care of Wesley and, despite my misgivings about leaving him with a babysitter, I realized that by this time in the wild, at a year and three months, he would have become a lone owl, no longer dependent on a parent. Happily, in spite of his shyness toward other people, Wesley allowed Wendy to feed him without incident. Somehow he understood that he was in her care.

The surgery went well and I came home, but the doctor's orders were "two months of rest and relaxation." At first this sentence was an irritating interruption of my life and work at Caltech.

"What am I going to do now, Wesley?" I said, stroking his neck.

He gazed back at me and then I realized: nothing. I couldn't go anywhere and I had no obligations at all. For the first time in my busy life, I could rest and, better yet, I had this lovely creature right here in front of me, and all the time in the world to spend with him. So I sat in bed or on the couch by the window with him standing on my head or shoulder and we just watched the world go by. At night I tethered Wesley to his perch next to my bed. Though he was older now, he still spent hours playing "helicopter" and doing mock perch attacks, always accompanied by rowdy screeches of joy. I found myself turning to Wesley and saying, "I'm going to go to sleep now, okay, Wesley? Go to sleep. Mommy goes to sleep." While it may not have been very scientific to refer to myself as "Mommy," I had had to choose one word to call myself, a consistent name that he could learn. Mommy seemed the obvious choice when he was little, and it stuck. Wesley would watch me turn over in bed and promptly go to sleep night after night. It wasn't long before I could say, "Go to sleep," and he'd slowly close his eyes, draw up one foot, and fluff his feathers as if he, too, were going to sleep. I say "as if" because I often caught Wesley peeking with one eye opened after a few minutes, only pretending. It was remarkable that he would try to emulate me because he was nocturnal. Even though he tried to be like me and sleep when I did, his natural instincts took over and he often woke me up at night. Eventually I could say "Go to sleep, Wesley" in the middle of the day and most of the time he would, which was handy when I needed a nap—and I often did.

By a few months old, Wesley had started to respond to many of the specific words and phrases I commonly used. I was not training him; he was learning naturally, the way a child would absorb the meaning of words as they are repeated over and over in the same circumstances or to describe the same behaviors. "Wesley you are sooo handsome. Mommy thinks you are sooo handsome," I would croon, always stroking his cheek feathers and kissing his nose as I did so. At some point along the way, he learned to associate these words with physical affection. Now I could say "You are sooo handsome" from across the room and he immediately lowered his nictitating eyelids, turned his face coyly to the side, and preened a feather or two, acting shy, until I came over and stroked his face.

Wesley loved to be told he was handsome. So for the rare occasions when I asked Wendy or my mother to watch him, I taught them to say, "You're so handsome" or "Go to sleep" to reassure him if he started acting up, and to say, "Here's your mice" when they fed him. It always calmed him down when other people knew his special words. He got it. He understood them, and it worked.

Ever since he was an owlet, I'd been giving Wesley verbal explanations of everything I did. Now I could walk up to his perch say, "Wesley, do you want some mice?" If he was full and didn't want any, he'd turn his head away. If he was mildly hungry, he would stare at me and slowly lower his nictitating eyelids. If he was very hungry, he would lunge toward me and holler about it, sometimes making begging sounds, which sounded like a handsaw being pulled across metal.

Eventually, Wesley's responses became more complex. He could answer a whole series of questions with his version of yes: lowering of the eyelids, direct eye contact, and sometimes excited audible responses; or no: turning his head away and not looking at me. We eventually developed a system much like twenty questions. "Wesley, do you want a mice?" (Whether there was one mouse or two mice, to be consistent, I always used the same word: "mice." I didn't expect him to learn plural versus singular.) Wesley would turn his head away if not hungry. "Okay, Wesley, do you want to go outside the room?" He would turn his head away in a no. "Wesley, do you want to play with water?" Head would turn away again; also a no. "Wesley, do you want a cuddle?" Instantly his head would whip around to make intense eye contact with me and he'd screech softly. Ah. That's it. He wants to cuddle.

Wesley learned this system on his own. I had not made it a mission to teach him to communicate with me as, say, biologists do with parrots or chimpanzees. And, from a strictly scientific point of view, Wesley wasn't speaking a vocal language. I do think that some of his consistent actions could be considered a primitive form of sign language, as well as his consistent sounds, if language is defined as a system of communication that works consistently between two sentient beings.

Clearly, Wesley had adapted beautifully to his unique situation living in a human, indoor environment. And he was communicating. While I won't claim that all barn owls have rudimentary language skills, I will claim that this one barn owl developed a rudimentary language by using his massive auditory cortex to adapt to an environment where this skill would be useful to him. Because I had talked to Wesley so much from his infancy, as his auditory brain developed, I think that he made the connection first between actions and words and then between actions and particular combinations of words like "It's time to go to sleep."

People, too, can understand another language without being able to speak it. One of my best friends grew up in a Japanese home where she and her siblings spoke only English and their parents spoke only Japanese. The parents understood English and the children understood Japanese, but neither could reproduce the language that they easily understood.

Similarly, it didn't matter that Wesley, given his vocal apparatus, couldn't make human sounds any more than it mattered that I couldn't make owl sounds. We developed a way of communicating with each other through our own language and gestures.

At two years of age, Wesley began to adapt his natural owl vocalizations to make new sounds to mean a variety of things. He adapted his begging sound, for example, to have slight variations in pitch, length, and intensity. Each new vocalization meant he was begging for a specific item. One variation meant "I want you to open the door." Another meant "I want water." Yet another one meant "Let me off my perch." Just the begging sound alone had about twenty new variations. I learned his meanings just as he had learned mine.

If I wanted him to land on my arm, I'd say, "Wesley, come here," and hold out my arm, tapping the spot where he should land. He'd fly to me immediately. But if he were uneasy about the landing, because my skin rolled around over the muscle, making an unstable place to land, he'd shake his head midair, and fly to a nearby spot instead, then wait for me to pick him up. When I sang or played music in the room, he would shake his head as if it hurt his ears. In fact, it probably did. With his sensitive hearing, my music must have seemed unbearably loud. I moved my electric piano out of the bedroom and began listening to CDs on headphones.

But I still noticed him shaking his head fairly often as if his ears were irritated even after I stopped playing music in the room. I was worried about his ears and for once, completely stumped. I called Dr. Penfield.

"I'm worried he's got an ear infection," I said.

"Oh, that's not likely. Was he unhappy, uncertain, bothered about something? Did you bring anything new into the room that might have upset him?"

"Yeah, actually. I put up some new pictures. But what would that have to do with his ears?"

"It's not that his ears are bothering him," said Dr. Penfield. "Barn owls do this when they are uncertain, irritated, or somewhat upset. It can be a little thing that bothers them, not enough to elicit a hiss. The head shaking is almost an unconscious gesture. But that's what it means. Something as simple as new artwork on the wall might be enough to trigger this response."

Almost invariably, Dr. Penfield knew exactly what an owl was trying to communicate. He taught me to observe carefully and to notice details that even many experts would miss. Because he modeled the value of attention to very subtle differences in behavior, I was able to observe that Wesley was changing his vocalizations significantly. He continued to alter them from normal barn owl sounds to new variations modified to fit specific scenarios. My childhood training in music was helpful, since I was sensitive to pitch, tone, and rhythm. I could tell when Wesley made a slight variation to a sound he had used in general situations, and I'd make note of it. This new variation would inevitably pertain only to a particular situation, for instance, if Wesley were reacting to my opening a window or answering the phone. Over time it became easier for me to know exactly what he was referring to. My nineteen years of living with Wesley gave me the unique opportunity to learn to communicate with an owl, which had never before been documented to such an extent. The great biologist Bernd Heinrich noted in his book *One Man's Owl* that his great horned owl was just starting to specialize his vocalizations at about two years. However, Dr. Heinrich kept his owl for only two years because he was sucessful in teaching it to survive in the wild.

An all-or-nothing creature, Wesley's emotional responses were as transparent as a child's. Things were black or white, good or bad, safe or dangerous, so he always expected me to follow through with exactly what I told him I was going to do.

Otherwise he'd become visibly upset, refuse to make eye contact, or even screech in protest until I did whatever I promised. Owls do not tolerate lies. If I said, "I'll play with you in two hours, Wesley," and then went about my business, two hours later he would start screeching and become unmanageable. I was stunned to learn that he had processed what *two hours* meant. I would say "in two hours" and then follow up two hours later, and did that so often that he was eventually able to figure out that it meant he had to wait a certain period of time. Somehow, he did know, generally, about how long two hours was, although he didn't learn other time periods in chunks of hours. He also learned what "tonight" meant, as well as "tomorrow." He may have been tapping into certain patterns of mine, but still, he was able to put those patterns together with my statements about "tonight" and "tomorrow." I was unable to discern how he knew all that he knew. I wouldn't have been as surprised if a dog had learned what Wesley did, but I hadn't expected an owl to have such abilities. Wesley would screech all through the night if I did not keep a commitment, making it impossible for me to sleep. It was just one of the many life lessons Wesley was teaching me, another Way of the Owl: if you make a promise, keep it.

One evening, I noticed Wesley chewing on the edge of a book, fascinated by the feel of the soft, slightly crispy pages. Taking away the book, I gave him a magazine, which he explored tentatively at first, but then began to rip into long shreds. He loved it! From then on, one of Wesley's favorite pastimes was tearing up magazines. I'd put one on his perch and he'd tear it gleefully into strips and piles of tiny beak-sized triangles onto which he'd pounce with abandon, like a kid in a pile of leaves. I'd say, "Do you want a magazine?" and Wesley would beg and jump around excitedly until I produced one.

Wesley would never get my old *Audubon* magazines, but after I finished *People*, *Rolling Stone*, or a catalog, they became owl fodder. He didn't seem to have a preference, though thicker magazines provided more of a challenge. This ritual eventually developed into "magazine night" as in "Do you want to have magazine night?" which he understood was entirely different from just, "Do you want a magazine?"

Magazine night meant we got ready for bed, then I gathered a whole pile of magazines and propped myself up in bed to read them while he flew around the room and pounced on things. After finishing a magazine I would put it on a pile next to his pillow on the bed, and he'd eventually fly down and start shredding the whole stack. He would sometimes rip and tear and mince those magazines for a couple of hours. Finally satis-

Wesley lies across Stacey's arm with feet up like landing gear. *Wendy Francisco.*

fied, he'd climb up on my lap and ask to be cuddled. I would lift him and lay him over my left arm, with his feet dangling and his head in my left hand, and I'd groom him while he slept.

While I was recuperating, I'd lie in bed and groom Wesley for hours. Wild owl mates do this for each other and it's very soothing to them. I found it soothing, too. With Wesley draped across my arm, I carefully groomed him, plucking off waxy tubes and smoothing new feathers out in proper alignment. He'd relax in my arms, making a gentle nibbling motion with his beak, which I'd try to imitate with the tips of my fingers while grooming. Wesley would reciprocate, running his beak lightly over my face and hair. But my hair was so long that he would lose his place and get tangled, so finally I just let him groom my bangs and he was happy with that. Wesley and I shared many such tranquil moments. We'd watch the cats, dogs, and goats playing in the backyard as mockingbirds dive-bombed their backsides trying to grab fur, clouds floating along in the sky on a windy day, leaves and birds flying past the window.

During one of these lazy afternoons, Wes began to get restless and beg for more mice. That was a bit unusual because he'd already had his quota for that day. But the four mice didn't seem to satisfy him and for the next few days, Wesley's appetite increased until he was eating seven mice a day and would not be happy with any less. I was flummoxed. I hadn't seen him eat like that since he was a baby. This continued for a few weeks and then I began to notice more feathers on the carpet than usual. They were everywhere. He was dropping them fast. One morning in July he flapped his wings vigorously, and the feathers lying all over the place blew up into the air. My room looked like the aftermath of a big pillow fight.

I thought Wesley must be sick, but he acted healthy and

full of energy. I fed him the requisite seven mice and called Dr. Penfield to ask what might be going on.

"Oh! He's molting. He must be two years old now. This is perfectly normal. My owl used to molt every year starting when he was two. It's very dramatic, isn't it?"

It certainly was. I was inundated with feathers. How could he lose so many and still have enough to fly and stay warm? Birds are automatically geared to respond to changes in the length of day, which triggers many of their cycles. The owls at Caltech didn't molt this suddenly because our lighting system was on a steady twelve-hour timer. So Wesley's extreme molt was a surprise for me. And I was amazed at the sheer variety of his feathers. There were so many different structures and types that they didn't even look like they came from the same bird. There were thick, fuzzy down feathers for warmth and insulation, and slick, curved body feathers for aerodynamics and flying. There were tiny facial feathers that looked like sparse Christmas trees; they helped to funnel sound toward his ears. But my favorites were the long, beautiful flight feathers, cream and gold with brown bars, built to carry him on the air. I gathered and kept them in a big box.

Wes had a lot of new pinfeathers growing in, so I had to be extra gentle when I handled him. Ever since he was a baby, Wesley emitted a high-pitched cry when I accidentally poked a pinfeather, which hurt him. Hearing his cry, I'd instinctively let go of him, so it wasn't too long before he began to figure out that if he made that sound I would release him instantly, and he started using it any time he wanted to be let down, even if he wasn't hurt.

Wesley was gentle with me, too. He had somehow learned not to scratch my skin and would lift his toes just enough so that the sharp tips of his talons didn't hurt me. He figured this out on his own even though I had never exclaimed in pain when he'd scratched me because I knew it would make him feel ashamed.

From two years old on he had a full molt every July. About a third of his feathers would fall out and a big rush of new feathers would grow back in. It was quite an event. He'd eat seven mice per day for over a month to prepare. Wesley let me know that he needed at least seven by begging incessantly until I gave him the right amount. Finally, I got the hang of it. By December he had finished replacing all of his feathers from the summer molt and was in full owl regalia. Right before Christmas, he had a secondary, less dramatic molt that was still a show. Every time he flapped his wings the whole room turned into a snow globe, feathers exploding in a flurry before gently settling down to the floor. Wesley's December molt was my California white Christmas.

9

Dr. Jekyll and Mr. Hyde

ACCORDING TO WILDLIFE rehabilitators and scientists, a tame owl like Wesley would never be able to survive in the wild because he could not develop the skills needed to hunt. Rehabilitators had learned this after releasing hundreds of barn owls they raised from babies back into the wild. All died within a short period not only because they couldn't hunt but also because they had no knowledge of what dangers to avoid—something their parents would have taught them. Since Wesley had failed at hunting mice when confronted with the one live mouse in the shower, I assumed that the pet finches I kept in my bedroom were perfectly safe. Wesley simply ignored them, using their cages as a perch when he was off his tether. Nonetheless, I always covered their cages when he was loose, just in case he was tempted.

So I was vaguely surprised when I came in one Friday afternoon after having let Wesley loose in the room to play on his own, to discover that one of the cages had fallen off the cupboard where it usually sat. Two finches were gone, and after a brief search, I couldn't find them. I became suspicious of Wesley that night when he wasn't particularly interested in his dinner. But I couldn't imagine he would have actually captured a pair of agile birds when he wasn't even able to hunt mice. Either he could hunt or he couldn't. I assumed that he had accidentally knocked the cage over and the finches had gotten out of the room on their own somehow. I moved all the other cages as far as possible from the edges of the cupboards on which they stood.

A week later it happened again, but this time there was evidence—a scattering of little feathers from one of my favorite finches on the floor. My heart sank. I was not pleased, but I didn't reproach Wesley, who had done it out of instinct. I couldn't blame a predator for being a predator, but I clearly needed to find new homes right away for my remaining fourteen birds. I decided to take them all to Caltech on Monday and see if anyone would take pity and adopt them. To get through the weekend without further incident, I kept Wesley leashed unless I was in the room to supervise him, even though it still didn't make any sense that he could have caught my birds.

Again I misjudged the situation, still picturing Wesley as a friendly, nonhunting owl—my little baby. That Sunday afternoon I took a nap when Wesley was off his perch. Crash! I woke just in time to see Wesley flying in circles around the room to gain speed, then pouncing hard against the side of one of the ten remaining birdcages. After he pounced on the cage of a zebra finch and pushed it off the edge of the cupboard, I jumped out of bed to rescue the little bird. But I was too late. The door had popped open on impact and the zebra finch had flown out in a panic. Wesley chased it with the most amazing flying I'd ever seen, matching that finch's intricate flight move for move. Darting up into a corner, flipping to go straight down, hugging the wall, zigzagging across the room with Wesley right on its tail, this little finch maneuvered so swiftly that Wesley looked huge and cumbersome in comparison. Yet Wesley never lost his rhythm. Finally, still in flight, Wesley reached forward with his feet like an eagle and grabbed the finch, falling to the ground with it in his talons. He had killed that little bird before I could get to him.

Wesley let out a victory cry. I was stunned. What had happened to my timid little guy who had cringed against the shower door in fear of a mouse? He was transformed into this aggressive, Red Baron, dogfighting machine—focused, unstoppable, and unbelievably capable. I was confused and upset about my finches, yet impressed by Wesley's abilities, which he had developed despite being raised in captivity.

While I kept Wesley tethered for the rest of the weekend, I did a lot of thinking. Would Wesley be able to hunt mice, now that he'd figured out how to hunt birds? We had been taught that, without exception, barn owls in the wild eat a diet consisting almost entirely of mice, with only 2 to 3 percent being small birds. Did Wesley have it all backwards? Would he have been a bird-eating barn owl in the wild? And why? He had expended so much energy catching that finch that, nutritionally, it almost wasn't worth the trouble. Predator survival is all about expending less energy to hunt than the energy gained from the food itself. Swooping down on a mouse is much easier than chasing a bird and provides the owl with many more calories.

I loaded nine cages—some holding pairs and some with single birds—into my car Monday morning and took them with me to Caltech. I would miss my pretty little birds and their songs and tweets, but I sure wasn't willing to risk any more being killed. One particularly softhearted woman adopted almost all of them on the spot. I was so grateful to her. Most Caltech postdocs live in large houses in Pasadena that they share with quite a few other postdocs, and I wondered what on earth her housemates thought when she came home with all these new pets. Of course, anyone who rooms with a biologist should expect these things.

No one knew the answers to my new questions about Wesley's hunting, but in spite of my eyewitness reports, the scientists were adamant that a barn owl raised by and imprinted on a human could never become a good hunter. Many times in the past, people had released owls who were unable to survive on their own—so much so that the birds trusted and approached humans to beg for food, only to be killed. Ignorant of the experts' beliefs about his abilities, however, Wesley was doing quite a few unexpected things, and my conception of his capabilities was broadening.

On Wesley's third birthday, I allowed him another opportunity to dispatch his own mouse. Even though I knew Wesley did not have the stamina to provide for himself in the wild, I wanted to see if he could learn to hunt mice without having observed an adult owl do so. In rehab centers, owls are raised from the egg to hunt. But to avoid having the owls associate humans with food or imprint on them, the rehabbers use puppets that look like adult owls to feed them from day one. Later the owls can learn to chase mice and hunt. Even though tame owls supposedly could not hunt, Wesley had certainly caught that finch midair as if he'd been doing it all of his life. Perhaps he would fool the experts again. I placed a live mouse in the shower and brought in Wesley.

The next three hours were so boring that I finally brought food for the mouse, who by now had explored the entire shower stall, drunk some drops of water near the drain, taken a nap, and washed his face. All the while, Wesley scurried respectfully out of his way. Every time the mouse moved, it would startle Wesley so badly that he'd jump up a half inch off the floor.

Wesley retreated to a corner of the shower and eyed the mouse. The mouse ignored Wesley altogether and turned its back to eat the food I'd provided. Wes gathered his courage and crept up behind the mouse, extended one toe, and touched it. Then he leaped backwards. Nothing happened, so Wesley again approached the mouse and picked it up carefully with his foot. The mouse bit him. Wes dropped it abruptly and ran back to the corner.

I gave up on our experiment and prepared to take Wesley back to his perch. But the mouse suddenly scuttled away from Wes, triggering some hidden instinct within him. Wesley pounced and broke the mouse's neck with his beak, killing it instantly. I was shocked and so was Wesley. He stood there for a moment with the mouse in his beak, looking startled, and then he realized what he'd done. He stood up straight and marched around the floor of the shower making loud cries of victory. I rolled my eyes. "As proud as Mommy is, I have to tell you, Wes, that no wild mouse in his right mind would wait three hours for a predator to decide to catch him."

Wesley hopped out of the shower and dashed behind the toilet with his mouse. He placed it on the floor, opened his wings and crouched over it. That was different. In the wild baby owls do this mantling to protect their food from their siblings. But I certainly wasn't any competition. "Wesley!" I chided, "I gave you that mouse." But he ignored me. When I tried to pick him up, he darted past me on foot with the mouse in his beak, galumphed back to the room, and flew up to his perch. I went to tether him as I usually did while he was eating so he wouldn't make a mess in the room, and he gave a scream of outrage and warning. That was different, too. Troubled by this unusual behavior, I left him to himself.

That night, when I tried to feed him the rest of his daily quota of mice, Wesley grabbed them from my hand and hunched over them, giving another scream of warning. He even lunged at me and snapped his beak. I knew it was a fake gesture, but still, he'd never done that to me before.

"Wesley, who just gave you those mice? I did!"

I wondered if I had created a monster. Was he turning wild? Did a young owl's newly acquired ability to hunt lead to the breaking of his bond with his parents? If so, why hadn't his success with the finch triggered these responses? Maybe he didn't see birds as real prey, or perhaps to Wesley finch hunting was just a form of play. And the most troubling question: Was Wesley going to change toward me?

For the next few days every feeding was the same hostile standoff. And all night long, Wesley flapped his wings hard, screamed a victory cry every half hour, and stomped loudly around his platform. By morning I was as exhausted as I had been when as a newborn he had needed feeding throughout the night.

Thankfully, this new behavior only lasted for three days following his mouse kill. Then he settled back to his normal self. I decided to allow him a live mouse once a year so that he could occasionally be his wild self. Every time, he reverted to the same jumpy, nervous owl he'd been the year before, afraid of the mouse until it made a sudden move and ran away. Once Wesley's instincts were triggered, he'd kill the mouse instantly and then go wild, treating me almost as if I were a stranger, but after three days he would return to his tame self again.

ABOUT SIX MONTHS after Wesley's third birthday there was a strange turn of events. One evening while Wendy, Annie, and I were watching TV in the living room, the most unearthly sound rose up around us. It was so loud we couldn't tell where it was coming from. It was so strange we couldn't even begin to guess what it was. Wendy and I looked at each other.

"What in the world is that?"

The noise vibrated the very foundations of the house; it was as if a huge UFO were descending upon us from outer space. Wendy ran outside to investigate and I ran back to my bedroom to check on Wesley, but he was nowhere to be found.

That deafening sound. What on earth was it? And where was Wesley? I searched frantically through my room: under the bed, behind furniture, and all around the drapes.

The closet. The noise was definitely louder near there. In fact, as I crawled inside, the sound was deafening. Adjusting my eyes to the dim light I spotted Wesley in the far back corner. Thank goodness! He was okay. I felt weak with relief, but that feeling didn't last long. Something was wrong. The unearthly din was coming from Wesley.

Stamping his feet up and down rhythmically as if marching in place, Wesley looked dazed, like a zombie, and he didn't seem to notice me.

"Wesley? Wesley?" I said.

No response.

"Wesley! What's wrong?"

I reached over to pick him up, but he pulled away, then grabbed on to my arm and started climbing. Was he sick, was he having some neurological problem? Oh, God, what was wrong with my little guy?

I backed out of the closet on my knees with him still clinging awkwardly to my arm. Every time I tried to reach for him with my other hand, he grabbed it with his beak and my wrist with his foot. Had he hit his head?

"Wesley, calm down. Just let me check you."

He didn't acknowledge me or listen to me at all. Again, he grabbed my hand with his beak and talons and pulled himself onto my arm, but then his body sagged as if he couldn't hold himself up.

He must have developed a muscular weakness—a classic symptom of stroke, aneurysm, or embolism. Or had he been poisoned? What had I overlooked in the room that might be killing him? I racked my brain as he struggled. Wendy had come to my room and watched solemnly from the doorway.

Suddenly Wesley stopped making the repetitive ear-splitting noise and started squawking like a parrot. I had never heard an owl make anything like that sound. This was getting more and more frightening. In apparent pain, he squawked and his body convulsed. He shuffled around and grabbed my arm between his knees while holding onto my hand with his beak. With every convulsion, his knees gripped my arm.

Maybe he had epilepsy? I'd once had a dog with grand mal seizures. The convulsions continued. With one last shuddering spasm he threw his head back and gave a great cry of pain and . . . then . . . seemed to return to normal.

It was over. He was okay. He flew up to his perch and started preening himself.

Wendy started to chuckle.

I was stupefied. Then I noticed a small drop of white fluid on my arm and finally realized what had just happened. My owl had just consummated his commitment to me—on my arm.

Wendy staggered down the hallway doubled over with laughter.

I slowly got up and went over to my microscope, pulled out

a slide, and scraped the droplet off of my arm. I flicked the light on and focused the scope. There they were—energetic strands of barn owl life force, racing for the goal.

Wesley continued to groom, looking quite satisfied with himself.

I didn't know whether to take a shower or have a cigarette. After I'd collected my thoughts, I tethered Wesley and followed Wendy into the kitchen for a cup of tea.

I had not realized Wesley was trying to mate with my arm because, for most birds, including owls, their reproductive organs are neatly and internally tucked away. Males don't have a penis and manage to fertilize the female by rubbing their cloaca (the opening under the tail) against hers to transfer a tiny drop of sperm onto the surface. Then the sperm swims up a long tube to fertilize the egg. It is all very tidy.

At three and a half years old, Wesley had reached the age at which barn owls mature sexually. I knew this theoretically, but had no idea that I would become the object of his affections. None of my other animals had wanted to marry me. But Wesley belonged to a solitary species that mated for life. Perhaps if Wesley had not chosen me as his mate, he would have grown distant from me and seen me as an adversary, as he did when he'd killed his first mouse. That's the Way of the Owl.

Now Wesley only had eyes for me and became quite serious about his responsibilities toward me. He grew very protective and fussed over me. He constantly sought out dark corners and little hidden spots, and tried to lure me to them with his ear-splitting nesting or mating call. One of his favorite places was the space behind the toilet, the perfect babe lair. He would drag magazines back there and rip them into fluffy nests. Barn owls are thought not to make nests, but Wesley certainly did. Perhaps some barn owls use nearby materials to create cushioning. We noticed at Caltech that some barn owls loved to rip up the coverings on their perches, whereas others ignored this material altogether. Wesley also filled the bathroom cupboards with magazine strips and called me to them. I always raced to find him when he did his nesting cry because it was so loud it shook the whole house, as if a stereo were on full blast. The only way to make Wesley stop was to let him mate with my arm. Then he would be quiet and docile, for a while. I wondered how long this nesting and mating phase was going to last. Fortunately, as long as I remained in the room with Wesley he didn't feel the need to call for me nonstop, but dealing with this demanding new dynamic wore me out.

One afternoon as I took a nap, Wesley was sleeping on his little pillow next to me. In my dreams I felt something soft brushing my face. Next thing I knew a mouse was neatly dropped into my mouth. Wesley was feeding me! Ugh, the pungent, skunk-like taste of rodent filled my senses. Ptuiy! I spat the mouse out and ran to the bathroom to gargle with Listerine.

Undeterred, Wesley flew after me with the mouse in his beak. He landed on the counter and began lunging at my face, screeching urgently. "Eat! Eat!" he was saying. I tried to avoid him but wherever I went, he followed me. Finally, he landed on my shoulder and leaned around, trying to force the mouse into my mouth.

"Okay, okay, Wesley, I'll eat it." I said.

I took the mouse from his beak and pretended to eat it, then hid it in my hand. No good. He went right to my hand and tried to pull it back out. He wasn't going to be fooled that easily. I showed him the mouse again and then turned my back and made noisy "yum-yum" sounds while I hid the mouse up my sleeve. I turned around and said, "All gone! Wow, that was great, Wes!

Thank you! Good boy!" He looked carefully for the mouse, bobbing his head from side to side, up and down, and then upside down. He finally decided that I had eaten it. His relief was obvious. His body relaxed and he flew back to his perch.

And so it was that I had to pretend to eat mice almost every day or he would become upset and worried that I was starving. Eating my own, human food in front of him did not impress him. No, it had to be mice and mice alone. As appetizing as Wesley seemed to find them, I was never tempted to eat mice, raw, steamed, boiled, or baked.

The actual owl mating and nesting season is from spring into summer. Sometimes a pair raises two nests of babies in a row. I believe that owls mate as a greeting and an expression of bonding throughout the year, even though the female may be fertile only during the nesting season. Wesley often did not produce sperm, so perhaps the male is fertile only during nesting season.

It was a relief to find out that other people who have cared for imprinted raptors had experienced this mating behavior. In fact, I heard of one guy whose endangered bird regularly mated with his hat. The scientist carefully collected the sperm and sent it to conservationists in breeding programs all over the world who were trying to artificially inseminate females to keep the species from going extinct. This bird's sperm was like gold to them. I didn't feel quite so weird after hearing this. Also, it is not unusual for animals to use some form of their mating ritual as a greeting ritual as well. Even humans and primates hug and kiss—often a part of mating—to greet each other, so it's not surprising that Wesley used his mating ritual to greet me.

Wesley was passionately attentive to me now in many other ways. He tried to communicate with me more adamantly and urgently than ever and his concern for me was evident.

Around this same time, I was dating a guy who shared my interest in songwriting. One afternoon, he came over to write music with me. I had recently discovered that, if I took Wesley off his perch and let him sit on a high piece of furniture, he would allow friends to come into the room as long as I stayed with them. Usually he would just nap with an occasional peek to make sure we were behaving properly. Norm and I were sitting in my room, playing guitar, when I mentioned to him that my closet door had come off its track and I couldn't lift it back on my own. Could he help me? Sure, he could. We positioned ourselves on either side of the door and said, "One, two, three, go" and hefted it back up into the track with a loud *pop*. Before I knew what happened, there was a loud scream and a blur of white and gold feathers. Norman was doubled over holding his face in his hands and Wesley had returned to his spot on the furniture.

Norm was yelling unintelligibly.

"Let me see," I insisted. I was worried that Wesley might have gotten him in the eyes.

He stood up and asked, "How bad is it, how bad? Tell me the truth, I can handle it."

There were just a few minor scratches right at the hairline above his forehead. I was relieved that it was nothing serious.

I looked over at Wesley, who was facing the wall looking ashamed. He had acted on instinct but now felt terrible for attacking a human being. I was amazed that he had tried to defend me. Little Wesley against a big human. What a hero, protecting me against all odds! He didn't know that the loud pop was just the closet door. He thought Norm had been hurting me and he was willing to lay down his life for me.

I said to Norm, "It's not serious, you'll be fine," and then went over to comfort Wesley.

"Oh Wesley, it's okay! It's okay; don't be upset! Thank you for trying to protect me."

Norman was getting pretty worked up on the other side of the room.

"Oh! You're glad he attacked me, then, is that it?"

"No, no! Of course not. Wesley thought you were trying to hurt me, don't you see? He is very upset, and I'm just trying to calm him."

"Oh, I see all right. I see that you care more about that owl than you do about me, bleeding to death over here. You should be worried about me, not that damn bird!"

Damn bird?

There was absolutely nothing I could do to redeem the situation. With sudden clarity I saw that Norm and I had no chance as a couple and that I wasn't the least bit upset about it. I started giggling. Norm just got more and more angry, then stomped out of the house, slamming the door and leaving behind his guitar. I gathered up Wesley, cuddled him, marveling at his loyalty and feelings for me.

Norm broke up with me a week later.

NOT LONG AFTER that, I went to work and noticed that everyone was in a somber mood. A few people's eyes were red-rimmed and they all looked exhausted. In our sunlit conference room I asked my supervisor what was going on.

"Oh, it's a terrible thing," he said. "Another lab has been using unreleasable vultures to study things in the environment. It's part of a bigger effort to figure out how to help them breed in captivity."

"Why?"

"They're getting ready to try for a big condor save. The

California condors are in such danger of extinction that they're thinking of trapping them all so they can protect them and trying to breed them in captivity. Then they'll teach the larger population how to survive in the wild. It's risky; if they don't breed, then the whole species will be extinct in a few years."

"So they're learning how to save the condors using vultures?" I asked.

"Yeah, no one wants to make any mistakes on the condors. Vultures are similar to condors so the scientists want to be rock solid in knowing exactly what their needs are. The lab had nine vultures."

"Had?"

"Yeah. Two days ago people broke into their lab, stole the vultures, and let them go, then trashed the lab. I guess these people have some beef with any animal being in captivity. They just took it upon themselves to let them go. Obviously, they know nothing about what it takes for the birds to survive out there."

"Oh no, what happened to the vultures?"

"A call was put out to other labs asking for help trying to find them. People have been looking for the last two days. Some of the birds were found, but they were so hurt they had to be put down. These were unreleasable vultures. Some were even missing a wing. It's one of the worst cases of abuse we've seen in a long time."

"That's hard to believe," I said.

"Yep, it's crazy, but some people don't even think dogs should live with people."

"What?" I was incredulous. "Do they think everyone is supposed to abandon their dogs by the side of the road?"

"Yeah, pretty much. There was a dog show where people ran in and let dogs out of cages, opened the main doors and yelled, 'Go free, go free.' Some of those dogs ran out into traffic and were hit. I'm all for improving conditions for animals, but this is some kind of wacko movement. I'm thinking they'll start going into people's houses and letting out their pets next."*

That statement stopped me cold. At this point in my life I couldn't even begin to imagine anything worse than losing Wesley. It had not occurred to me that it would be risky for people to know about him.

A few nights later I dreamed that someone had found out about Wesley and let him out, yelling, "Fly free, fly free!" Wesley flew in a panic to a high tree and then circled out farther and farther away, startled by all the new sights and sounds, narrowly missing danger after danger. I was trying desperately to catch up to him, calling him, but he was too confused and frightened to respond. I watched helplessly as he flew over a freeway and into the cars. I woke up in a cold sweat, my heart pounding. Wesley was right next to me, sitting peacefully on his perch. I turned on the light, picked him up, and cuddled him until I calmed down.

I decided that from then on I would tell only trusted friends about him. But, unfortunately, Wesley was advertising his own presence with his interminable mating call. Every time he did it I thought, *Oh, no, people can hear him from the street.* But in all the years Wesley was with me, no one ever commented. Even when guests sat at the dining room table having tea while Wesley wailed away down the hall, they just raised their voices and continued conversing over the din. It was amazing. I even asked a few people later on what they thought the sound was. They had to think about it, and if they did remember, they said they thought it was a very loud dryer/fan/air conditioner in desperate need of oil.

---

* I took this story at face value at the time and have not been able to confirm whether or not the incident happened or happened in the way it was described.

What we do not expect we often do not even perceive. It would not have occurred to visitors that I might be raising a real live owl in the house. Normal people didn't do such things. But in my line of work, many people I knew were far from normal, and my lifestyle didn't even begin to compare with the way some other biologists had chosen to live. If I hadn't known them personally and been to their homes, I never would have believed it myself. I worked with many of these folks every day.

Nine-year-old Wesley looks at Stacey through the mirror on her makeup case. *Stacey O'Brien.*

10

A Day in the Life of a Biologist

I GOT UP particularly early one morning in order to stop by a marine biology lab on my way to work to pick up a couple of octopi for Tom, a postdoc at Caltech who was studying their behavior. I love octopi and was looking forward to seeing them.

I dressed in my usual sweatpants and T-shirt because it didn't make sense to get all gussied up to work with animals. I'd just wash my face and brush my teeth and wait till the end of the day to shower.

Wesley played in the water while I got ready and then settled down on his perch to sleep while I was gone. It was hard to leave him behind in the mornings, but I always gave him a snuggle and kissed him good-bye on the curve of his smooth, warm beak as I left.

After being stuck in Los Angeles traffic, barely moving, I finally reached my destination, the marine lab at Occidental College, which smelled like a hundred years' worth of fish and formaldehyde. I was surprised to find only one container waiting for me, because I had expected two. Octopi need to be kept separate, as some species will eat each other, but the scientists were already loading diving equipment into the vans, getting ready to head out to the research boat for the day.

I waved at one of my old Occidental classmates, Lisa, who was getting ready to go diving with the group. Lisa could walk right up to a bloated, decaying seal carcass on the beach, reach into the black, liquefying flesh up to her armpit, fish around in there, and pull out a bone. "Hmm," she'd ponder. "The bones were weak." Everyone else would be running away retching. This proclivity has helped her in her future career, as she has spent her life doing autopsies on dead sea mammals to figure out what's killing them off and how to try to protect the living—a job she was born to do.

"Why are the octopi in the same container?" I asked one of the lab assistants. "Won't they eat each other?"

"Well, it's only fifteen minutes to Caltech, so if you hurry it'll probably be fine. We just now put them in there," he replied.

I loaded the container into my car and looked inside. There, swimming in the seawater, was one huge octopus and one small one. Not good. I drove to Caltech as fast as I could and parked in the loading zone. The ice chest was heavy, but I lugged it out of the car and rushed toward the back door of the building.

"Stop right there! What are you doing?" I looked up and there was a stern cop standing in my way. "This is a loading zone," he barked.

"I'm unloading," I said.

He took out his ticket book.

"I'm unloading octopi," I insisted.

"Yeah, right."

"Listen," I blurted out, "I have two octopi in here. One is much bigger than the other and will eat the little one if I don't get them into the lab right away," and I yanked the top of the ice chest off.

To my dismay, the big one had already eaten the small one and was undulating around, happily changing colors, as they do to express emotions. He reached one long tentacle over the top edge of the container and began feeling his way out using his suction cups.

"Oh, no" I yelled loudly. "It's already happened! He ate the little guy. Now, what am I going to do? Look, there's only one pathetic little tentacle left floating on the—"

I looked up to see the cop stumbling backwards, making choking, guttural sounds. He jumped into his car and burned rubber out of the parking lot.

This is how stories get started about secret laboratories doing experiments on aliens.

I set the container near the lab door then went upstairs to get Tom. A wave of chemical odors mixed with the pungent aroma of animals hit me. I would always love that combination of smells. As we took the ice chest up the elevator I explained the tragic demise of the smaller octopus.

"That's a real loss," he said.

"I know, I should have insisted they separate them, but they thought it would be all right." At least we had this one, and into the aquarium he went, full and content.

I went off to start my workday, cleaning animal cages and feeding the owls and songbirds. It was also my responsibility to check all the animals for signs of ill health and other anomalies. It took about four hours to make my rounds, which gave me a lot of time to think.

I loved being at Caltech. It had been a part of my life since I was eight years old, but recently I had begun to worry about my financial future. There really wasn't any way to have a career at Caltech without a PhD. An aerospace company was recruiting me quite persistently and they were starting to turn my head with their offers. The position was out of my field but paid a lot by my standards, and they would train me to work with UNIX operating systems. The sky was the limit as a UNIX specialist, although biology would always be my first love. It would break my heart to leave, but I had spoken to Dr. Penfield about it and he had assured me that the lab would still welcome my observations of Wesley. In that way I'd still be involved with the lab, which kept me from feeling that I was losing everything.

I went back up to the main building for some supplies and dropped by the lab of John, a postdoc one floor down from us owl biologists. As was my habit, I leaned against the door of his small lab area to chat. I never went all the way inside for fear of bumping into one of the crowded shelves and bringing their contents down on my head. His lab had thousands of black widow spiders living in thousands of petri-like dishes that had been made into little spider habitats, stacked high on shelves that went up to the ceiling. I hated to think what would happen in an earthquake.

John particularly doted on his "nursery" of egg sacks and newly hatched babies. Clucking and fussing over them like a mother hen, he would separate the babies using a glass tube. He gently sucked one baby spider into the tube, transported it to a new dish, then carefully blew it out into its new home. I always feared the little spider would run right up into his mouth, but it never happened.

John spent long hours sorting out all the babies when they hatched, more hours than he spent with people. He thought it was sooo adorable when they hunched up, right before they jumped. I finally conceded that, if you watched them long enough, you could see their tiny faces, which had an alien sort of "cuteness" to them. He would say, "Oh, look, they're so precious when they're little!" And the babies were, in a strange way, delicate, perfect miniatures. Not lanky like the adults, but more, uh, cuddly looking.

At some point I think maybe John lost perspective. He started taking the black widows home. Then he got to where he refused to allow any yard work whatsoever because it disturbed the wild spiders, so his yard became completely overgrown and covered in huge, thick, active spider webs. Eventually he became a professor of biology at another prestigious university.

John was not married, although he was charming and gorgeous. What a waste. But I couldn't imagine dating a guy with this specialty. After all, how would you raise children in a home filled with black widow spiders?

So many interesting people from all over the world came through Caltech. Jane Goodall's lecture there changed my life when I was eight. Her discovery that chimps made and used tools shook the world of science to the core and showed how close these primate relatives are to us humans. To me, Dr. Goodall is the Galileo of behavioral biology. Every year for nearly ten years my sister, dad, and I went to hear her lectures whenever she came to Southern California.

Dr. Goodall's biggest influence on my life has been her refusal to see animals as simply instinct-driven, stimulus-response mechanisms, which has been the dominant view of many scientists since the overly reductionist Enlightenment scientist, Descartes, declared that animals had no real feelings, and his twentieth-century descendants, the behaviorists, led by B. F. Skinner, described them as little more than furry automatons. Goodall has proved that each chimp is a unique, sentient being, and other scientists have taken courage from her and have studied other animals, such as elephants, orangutans, gorillas, and wolves, proving that these creatures, too, exhibit individuality and personality. This encouraged me to go beyond just studying Wes, and to experience him as an authentic feeling, intelligent, and even spiritual individual. I was Wesley's friend, not a superior observing a specimen. The emphasis on empiricism in behavioral biology often keeps us from seeing clearly, and may actually bias and block us from observing the very truths we seek.

Many scientists who will never be as famous as Dr. Goodall make important contributions to science through their own particular specialties, which can seem quite odd to laypeople—even odder than studying black widow spiders. For instance, one of my favorite professors at Occidental College spent his entire life studying the ovary of the Pacific surf perch. That was his passion. As he got ready to retire, I used to joke to him that he had spent his life studying the right ovary of the surf perch, and now he would be free to study the left ovary (they're the same). When he did retire, he indeed moved to a house next to the ocean, set up a lab, and continued to study the ovary of the Pacific surf perch.

When I was a kid, my dad was friends with Richard Feynman before he'd won his Nobel Prize in physics. Always an iconoclast, Feynman never let anyone tell him how to act or behave. He would go to topless bars to sit there and do calculations on the tablecloth. He wasn't there to look at naked girls; he just liked the ambiance. Even after he won the Nobel Prize, he didn't let it go to his head but still taught freshman physics. He was so entertaining that he eventually wrote a popular science book that became an enormous bestseller, *Surely You're Joking, Mr. Feynman!* and his lectures still sell to physics students worldwide. His ability to see clearly, without bias, enabled him to demonstrate memorably to Congress why the O-ring shattered and caused the tragic *Challenger* space shuttle explosion in 1986.

Yet another physicist at Caltech insisted on working in the

buff in his office. There was a picture in one of the hallways in the physics department of him sitting naked at his desk, taken tastefully from the side. I used to see a man striding across campus decked out in an authentic-looking court jester's costume from the Middle Ages. He had the funny hat, striped bloomers, purple tights, and velvety purple shoes with the toe curling up over the top, decorated with a jingly bell. No one ever looked twice as he walked by. I don't know who he was, but people treated him with respect.

When I returned to the big barnlike owl building after my chat with John, I noticed that the adolescent wild owls, the most unruly of the bunch, who were housed together in one large aviary, were particularly restless that day because of the Santa Ana winds. Hot, desiccating winds that come in from the desert, the Santa Anas fill the air with static electricity and generally irritate everyone, including animals. I decided to play some soothing music so I tuned the upstairs radio to a soft classical station. Then I went in to clean the adolescents' aviary wearing a helmet, face shield, rubber boots, and my lab coat.

The birds finally settled down into their favorite perching areas and I started working. I didn't notice that the music had changed until a soprano started screaming at the climax of some opera. Terrified, the owls flew around frantically as the singer's voice topped out at the highest notes and the orchestra hit a sustained crescendo with crashing timpani. Dozens of owls were bumping into each other and smacking into me, tumbling to the ground. Several landed on me, then panicked even more when they realized they were standing on a human being. Some owls attacked me, hitting my shoulders and arms.

I slipped out the door and raced up the stairs to turn off the radio. The owls calmed down, some of them just sitting on the ground panting. I sat on the floor, weak in the knees, trembling and laughing a little hysterically, covered with dots of blood where the terrified owls had pierced my lab coat with their talons. I decided the place was clean enough for the day and finished up by feeding them, still shaking.

It was time for lunch, so I cleaned my wounds with a solution of iodine and went back to the office building. I avoided the cafeteria, aptly called "the Greasy." Like others, I tended to bring lunch from home and join my colleagues around the conference room table exchanging stories. As we ate, tame owls flew in and out whenever they wanted, checking on their humans and each other. None ever interrupted us with "Howlers," the disciplinary letters that owls delivered to Hogwarts students in the Harry Potter books.

After lunch I hurried down the hall and almost ran into Steve. This is not something you'd want to do physically because Steve's skin was infested with parasites. A member of the Caltech primate group, where I had worked before switching to the owl project, he studied owl monkeys, so named because they have faces like owls. Steve was a Jane Goodall–type field ethologist—a wild animal behavioral biologist—who went deep, deep into the Amazon jungle alone. Just getting there over land and by canoe took some six weeks. His research area was a dense swamp, so he would be up to his knees in water from the Amazon, which was a stew of parasites. All night, every night, he watched the owl monkeys up in the trees, which was not good for his neck. His boots would rot off his feet within the first month. After his boots rotted off, his feet rotted, too. Steve slowly became part of the jungle itself, a human host for the parasites and hangers-on that the Amazon had to offer.

I don't know how long he was there, but it was a long

time. Long enough to become kind of "Amazon-ed." Changed. Altered by the experience. Not one of the regular folks anymore, if you know what I mean. He had a different outlook on life. Steve had so many different kinds of parasites that I hadn't even heard of many of them (and I had studied quite a few). The most impressive was a creature that laid eggs right in his flesh. The worms that emerged from these eggs grew to maturity just under his skin. You'd be standing there trying to have a straightforward conversation with the guy with worms moving under his skin. Steve just left those worms alone until they were ready to "hatch" to the next level in their life cycle—in which they poked through his skin. He would know when they were ready because they would make the spot itch so that he would scratch and create a little opening for this fat 2-inch-long worm to crawl out. Sometimes the worm would poke its head out and when he tried to grab it, the thing would dodge back into the hole in his skin. So he had to almost catch them off guard. He did this quite matter-of-factly, as if everyone in the world had worms crawling out of his skin. To him it wasn't an issue. No problem.

"Carry on, what were you saying now?" he'd say as he rummaged around for a jar. He'd save the worms he hatched for doctors who specialized in tropical diseases, to whom these worms were real treasures, appearing as they did right here in Southern California. A human petri dish, Steve had attracted a whole troupe of tropical disease specialists who were thrilled to have the opportunity to study him. Sharing their scientific curiosity, he was very cooperative.

Steve told me about another postdoc who was going to be heading off into the jungles soon.

"Once he gets his appendix out he'll be good to go," he said.

"He has appendicitis?" I asked.

"No, no," he laughed. "Most people get their appendix out before going into remote areas, didn't you know that? Think about it. What would happen if you were six weeks away from help and you got appendicitis? You'd die. It's not worth the risk, so most people just have it out."

There are an awful lot of people studying in the Amazon, and these are the chances they take. No one at their institutions or universities is checking up on them to see if they need anything or are still alive—just "he should reappear in about a year's time, we think, but we have no idea where he is." These field biologists are so fascinated by the creature they are studying that they devote everything to their science.

Another female friend studied a more "domestic science," doing research at a medical center trying to develop a birth control pill for men. When they finally figure it out, we'll all hear the announcement, "Scientists have discovered a birth control pill for men," but almost no one will wonder or hear about who or what was actually involved in making this discovery.

My friend's work entailed studying the attributes of sperm as it worked its way through a sort of assembly line in the factory that is the testicle. Each spermatozoon starts out as a nonspecific cell and goes through some fourteen or so distinct stages of development before actually becoming a sperm cell. The researchers were trying to find a way to interrupt this process so that the sperm cell would not fully mature. That's the aim, but the actual work went like this:

Every day she was presented with a "bucket o' balls," the testicles of some of the John Does and other men who had donated their bodies to science. The first thing she did in the morning was empty the bucket into a large blender, much like the one you have at home for making smoothies. And that's what she made: a testicle smoothie that she would then run through a machine that separated the developing sperm cells according to fourteen or so types. Then she took the fourteen or so sperm soups and studied them and the effects of experimental medications upon them.

Scientists often find themselves in the most extraordinary situations and want to tell people about them. Yet most nonscientists are easily grossed out or simply aren't interested, which can be frustrating.

Let's say that you are a married entomologist (such luck!), and your husband asks you to pick up the takeout order you placed. Off you go in the "good car," which is a rare treat in itself, because you drive a clunker to work. (You simply cannot see why anyone would need anything more than a clunker, just to get from point A to point B, or to transport your boxes of insects.) Off you go.

You get to the restaurant and, outside, after you've picked up the food, you see the most amazing thing on the restaurant's porch. To others, it would look like two bugs, but you see that it's a fight between a carpenter ant and a fly. You stop to watch for just a little while. You drive home with the food and can't wait to tell your mate the whole story. You've really built up a head of steam over this thing and you're eager to describe every detail. You pull in and race into the house all excited.

Your mate is oddly cold and distant. He says, "Why did it take you two hours to pick up food from a restaurant one mile away?"

"Oh!" you answer. "Just listen to this. I saw the most extraordinary fight between a carpenter ant and a fly. I am not kidding you! I think it's unprecedented. And you're not going to believe this—the fly started it!"

"Just gimme the damn food," he snaps, grabs it out of your hand, and stomps into the kitchen.

You hear cupboard doors slamming, silverware being thrown forcefully onto the table, the microwave whirring to reheat your dinner. You sense that you might not be able to tell your story until you get to work tomorrow with the other scientists, who will understand completely. You lie awake all night rehearsing and rehashing the incident in your mind, trying to make sure you get all the details right for when you tell the gang. They won't roll their eyes.

And you are right. Not only do they share your enthusiasm, they press you for more details on exactly how did the ant come at the fly in the 123rd pass. Was it from the right? Did they appear to learn from trial and error? Was the fly on its back or did it rear up on its hind legs? How did each animal use its mouth parts, and to what end? How did they protect their antennae—pretty soon there are objects on the table and a group of scientists gathered around. This actually happened, by the way.

"Okay, now this eraser is the fly and the soda can is the ant . . ." The crowd around you grows as the news races throughout the offices.

"Go to the snack room! Someone is describing a fight between an ant and a fly. It's unprecedented!"

The other scientists drop everything and sprint to the snack room. Arriving breathless, they ask, "What'd we miss?" and there's a murmur as someone fills them in on what has happened so far. Excitement fills the room. You wonder why you don't get this reaction at home. What is wrong with your mate?

VISITING BIOLOGISTS FROM all around the world would come to Caltech for a year or so to study our work. Some of them were from cultures that did not hold with our American habits of hygiene. You would think that with our lack of concern about guts and animal smells, biologists would not be sensitive to human odors, but for some reason this is not so. In our lab the line was drawn at not bathing. We may be able to dig maggots out of the flesh of a living rescued animal to save it, while enduring a stench that would cause most people to pass out, but we retch at the scent of an unbathed person. Even those of us who worked with monkeys in an atmosphere reeking of primate often could not stand the smell of a stinking human.

Ignorance of the actual dynamics of daily life can be bliss sometimes. Because we know chemistry and biology, we knew that when we smelled something, the molecules from the source of the smell had actually entered our noses and taken up residence on our receptors. So when we smelled a dirty person, this meant that some of his filthy molecules had actually gotten into our nasal passages. This bothered us. We didn't want to know that person that well, and we certainly didn't want his disgusting molecules in our nasal receptors.

Gagging coworkers finally put up a protest, and our boss elected one of the supervisors to give "the talk" whenever an unbathed individual reported for duty. The talk said, basically, that the rules of the lab were that you had to shower thoroughly each and every day without fail, including washing your hair, and you had to use soap. And you had to wear freshly washed clothing every day, whether your clothing from the day before looked clean or not (we had learned that if there was no actual dirt on a garment, some scientists would wear it forever without washing it) and that included underwear. The visiting scientist also had to brush his teeth and use a deodorant and antiperspirant daily.

Some of the visitors were quite taken aback by these rules, but our scientists remained steadfast in their insistence that these standards be laid out and enforced by management. A sheet with full hygiene instructions was given out to each scientist. This was groundbreaking stuff for some folks and it seemed outrageous to them. But we stood our ground.

After my encounter with Steve, who was very clean other than his parasitical hangers-on, I was quite happy to go up to the labs, where all the freshly scrubbed biologists were working, and do some microsurgery under the tutelage of my supervisor. It was a real privilege to be doing such leading-edge work, including many intricate procedures, for example, to inject a tiny finch egg with a microscopic glass needle using a foot-controlled microscope. We would insert monoclonal antibody tracers into the developing embryo so we could track later which neurological cells were developing at the time of the injection. With careful record keeping, we could see how the brain developed in the embryo at each stage. After inserting the tracers, we had to reseal the egg with a tiny drop of candle wax, all while managing not to kill the embryo inside. Then we returned the eggs to their nests where their parents hatched them and they led normal lives. In another delicate procedure, we had to sex the finches, since we were keeping breeding pairs. Sexing involved threading a microscopic optical filament between two tiny ribs to look down into the sex organ area, which is inside the abdomen near the diaphragm below the lungs, to see whether a finch had ovaries or testicles. We always used full anesthesia—of course—which necessitated learning another odd technique, mouth-to-beak-resuscitation.

Anesthesia on a very small bird is tricky, and if a bird were accidentally injected with too much, it would stop breathing; then you had to be able to breathe for the bird until the anesthesia wore off, which, thankfully, it did quickly. Whenever a finch died of natural causes we would practice this technique in order to perfect it in case we needed it for the live finches. Right . . . My mouth on a dead bird's beak. Our motto was "No Waste, No Pain, No Harm," and we interpreted the "No Waste" part to include getting some use out of the dead finch. The trick was to blow carefully so as not to burst the lungs. I am proud to say that I never have burst a lung of any animal, not even a dead finch. I don't think we ever made an anesthesia mistake, but if we had, we were ready. That was Caltech—thorough.

We were also called upon to sex birds from other animal centers because our techniques were so advanced. If a zoo had an extremely endangered pair of ruffled toucans, for example, and they wanted to see if they were male and female so they could start breeding them, the caretakers came to us rather than to a vet. In those archaic days, twenty-some-odd years ago, vets would make a long incision and open up the bird like a book, killing it half the time, just to see what sex it was. Zoos and top breeders obviously preferred our microfilament techniques.

While absorbed in these procedures, my supervisor and I would chat about all kinds of things. One day, I decided to ask him if he knew other biologists whose work had affected their way of life.

"Oh, do I ever," he said. "Soon after the vulture lab was attacked and those poor birds released, there was another attack on a facility that was doing research on childhood leukemia. It had taken ten years to breed a mouse that was crucial to the study, and all those mice were released. That means children will die because the results of this research will have to wait another decade."

"That's horrible," I said.

"Yes. Scientists all over started getting worried about their own lab animals. A primatologist I know was so attached to his monkeys that he couldn't bear the thought of any harm coming to them during the night. He tried sleeping in his lab every night, but that couldn't last, so he started taking them home, one by one. Talk about a change in lifestyle. First, his very favorite monkey. Then another one that he just couldn't stand to lose. Then it snowballed. The last time I visited this guy he had laid down a foot and a half of sawdust throughout his house and there were fourteen monkeys living freely in the home. You could smell the place from half a mile away."

"What did the neighbors say?" I wondered.

"Fortunately, he lives way out in the country or he could never do this. The monkeys do have cages, but they're not used except for time-out."

"What's time-out?" I asked.

"Well, the monkeys are like children, they're so smart. And they love to torment his dogs by pulling their hair then leaping out of reach. The primatologist figured he'd have to teach them not to do this, so he established a rule where the monkey in trouble had to go sit in his cage for five minutes if he pulled the dogs' hair."

"Did it work?" I asked

"Only for a few days. Then the monkeys would just pull the dogs' hair and go straight to their cage and sit for five minutes without being asked. They figured it was part of the routine. It was no deterrence whatsoever." He laughed. "This guy has these monkeys sitting around the table in the morning eating toast and marmalade with the family."

"I hope he doesn't try to dress them up and make them act like humans," I said.

"Oh, no, no, he'd never do that. They have a great life, and the dogs are there to guard the place so that no one will mess with his precious monkeys."

"Wait—family? You mean he's married?" I asked.

"Why, you want his phone number?"

"No, I'm just wondering if people ever do find someone who will put up with the way they live."

"Oh, yeah, he's married. She's a primatologist, too."

Aha. So that was the key. You had to marry someone just as weird as you were. Hmm.

After I finished my work with my supervisor, I wandered down the halls to check on some of the animals. Suddenly a closet door opened right in front of me, and a furry man walked out. He was what we called a "troll." Unshaven, his beard and hair both reached his belt. He didn't appear to notice me at all. He shuffled down the hall and disappeared into one of the bathrooms.

Theoretical mathematicians and physicists, trolls are ubiquitous at Caltech and go as far back into its history as anyone can remember. Caltech was built in the 1800s and was heated with steam that ran through a labyrinth of tunnels with all kinds of twists and turns. The steam and hot water pipes still run through the tunnels, making them warm in winter and comfortable in the summer. The trolls live deep in the labyrinth, rarely coming aboveground. That is their home and it's okay with everyone. They receive grants and their meager style of living doesn't cost much.

Each building has secret doors in certain closets that lead into the labyrinths so the trolls can go from building to building and use the locker rooms. People say Caltech is as close to Hogwarts as one can get in the real world, and I'd have to agree. I've been down in those tunnels, and as I walked through the darkness, I'd occasionally come upon a bluish glow, the computer screen of a troll. Next to the computer screen, in a small alcove, would be a twin bed, some blankets, piles of books and papers, and the computer. That was it. Some of them live their entire lives this way. Productive genius theoreticians, they tend to keep to themselves and publish their work. Some of them clearly have what's now referred to as Asperger syndrome, a mild form of functional autism, but they are happy in their secret cubbyholes, doing calculations and making discoveries. After all, theoretical scientists do not require a lab—only a piece of paper, a pencil, and a fantastic brain.

MY WORKDAY OVER, I hit the LA traffic once more. Home at last, I was greeted by the usual chorus of geese and horses. Wendy was in the living room, playing her guitar, working on a song. Annie was drawing a crayon project on a big sketch pad. Omar, Wendy's umbrella cockatoo, sat on the back of a chair and bobbed his head to the beat, occasionally screaming, his idea of singing along.

As soon as Wesley heard my voice, he screeched and I went straight to him, as I always did, kissed him on the beak, cuddled a few minutes, then let him off his perch to play.

I had gotten several cages of Syrian ground squirrels, better known as teddy bear hamsters, after the disaster with the finches. The hamsters were much larger than finches and Wesley didn't see them as prey. In fact, he seemed to be entertained, seeing them as a sort of "owl television" brought in solely for his viewing pleasure. He didn't even appear to mind my playing with them. He assigned himself as their guard and would

Stacey kissing Wesley; Wesley nervous about the photographer. *Connie Fossa.*

screech to let me know if one somehow escaped its cage, so I would immediately come put it back in. As surprising as it was, we coexisted quite well. I needed multiple animals around me and loved the sound of Wesley and all the hamsters playing at night. The room thrummed with life and I felt like I was sleeping in a forest instead of a suburban bedroom.

That night, I stayed up late reading, and just before retiring checked the hamsters. One of them seemed to have died. Wait, she was just "mostly dead"—she wasn't breathing but still had an occasional heartbeat—maybe one per thirty seconds. Her temperature was normal, which meant she wasn't going into a dormant, or hibernating, state. I checked her air passage and it was clear, so, holding her in my right hand, her tongue out and secured by my thumb, I started doing mouth-to-mouth resuscitation. Wesley watched all of this quite calmly from his perch, where he was tethered.

Still breathing for the hamster, I ran with her into the kitchen, scribbled a note for Wendy, leapt into my car, and sped down the freeway toward an all-night exotic animal hospital forty-five minutes away, hamster in one hand, driving with the other. I started compressions with my right fingers, since her heart was barely beating and I needed to get the oxygen into her organs. I'd covered her entire face with my mouth, doing tiny puffs, peering over the wheel. Thank God for all the practice on finches.

I was weaving like a drunk driver, trying to focus on the hamster, the car, and drive with one hand, but there was no one on the road at 3:00 a.m. so I wasn't too worried. But suddenly I noticed lights and sirens behind me. Unbelievable. The second cop of the day. I had become a menace to society.

I stopped, rolled down the window, continuing resuscitation, and one of the two officers who'd come up to the car said, "Do you know you're driving like you're drunk?"

"Yeah, I realize that [*puff puff*]. I'm sorry [*puff puff*]. I'm doing CPR on a [*puff puff*] hamster and trying to get her to the all-night animal hospital!"

"You're what?"

Out came the flashlights. Seeing the hamster lying belly up in my hand, her head back, tongue out and held down by my thumb, they leaned into the car and watched as I did compressions and mouth to mouth. The cop hit the hood of my car.

"*Go go go!* I've never seen anything like this in my entire career! If that doesn't—"

I didn't hear the rest of what he said because I was already speeding down the freeway again. When I got to the hospital my hamster was in a seizure-induced coma of some kind. I continued resuscitation for a full hour until she cough, cough, coughed and started breathing on her own. She lived a long, productive life afterward. The vet there couldn't believe it. "Where the heck did you learn to do that?" he asked.

Well, I'm a biologist at this lab at Caltech . . .

11

Owls Are Not Waterbirds

I ACCEPTED THE offer at the Aerospace Corporation in El Segundo. It was difficult saying good-bye to everyone at Caltech. Working with computers would be a far cry from my daily adventures with animals and animal biologists.

On my last day Dr. Penfield loaned me some recording equipment. "I want you to record the sounds Wesley makes and then bring us the tapes," he told me. It felt good to have a last assignment and to know I would still be connected.

Wendy's home was too far from the Aerospace Corporation for me, so I sadly broke the news to her. We both had tears in our eyes.

"I won't be that far away," I said. "We can visit each other as much as we want."

But we both knew it wouldn't be the same and that this was the end of a wonderful period of our lives.

When I had told my mom, who lived in Huntington Beach, that I'd accepted the new job, she had said, "Oh, you ought to come live here with me."

"Even with Wesley?" I asked.

"Sure, if he doesn't poop all over the house."

"He wouldn't do a thing like that. After all, he is your grandson."

She groaned.

"So you really mean it, Mom?" I asked.

"Yes, you could save some money and get ahead a little. I've got two empty bedrooms, and it's a short bike ride to the beach. Move in as soon as you can."

When it was finally time to leave Wendy's place, I packed my car and put Wesley in his carrier. Leaving him in my now-empty bedroom for a few minutes, I went around to the backyard to give Courtney the dog a big hug. I said good-bye to the goats and horses. The geese honked in their normal loud chorus as I went back into the house.

Wendy was standing in the entryway with Annie at her side wearing an adorable calico dress. With long, thick, curly hair, she was a lovely child—a thoughtful, circumspect little girl with a wit that belied her youth. Finally, I went to get Wesley. Wendy leaned over and said good-bye to him through the carrier door. He seemed to sense the seriousness of the moment and didn't threaten her as usual, but chirruped and twittered instead, maybe because he was now out of his own territory.

I gave Wendy and Annie a big hug. "Good-bye! Good-bye! Keep in touch! Good luck!" and Wesley and I were on our way.

It was a long drive down the coast to my mom's house. Wesley slept peacefully in his carrier, having grown accustomed to this mode of travel from our long commutes to Caltech during his baby days. I drove past a "Welcome to Huntington Beach" sign and opened the window. A blast of cool salty ocean air filled the car.

My mom was waiting with a big hello. She leaned down to say hi to Wesley and he chirped right back. My bedroom was big and Wesley would have plenty of room to fly. As soon as he saw his perch go up he knew we were moving in together and all was well. He jumped up to his spot on the top dowel, groomed himself, fluffed up slowly to maximum poofiness, then shook himself head to toe like a dog. Now, he was settled in. Just like that. Some animals—especially cats—are upset by big changes, but this move didn't seem to bother Wesley at all.

I had a week before my new job started. I usually wore old sweats and T-shirts with lab coats to Caltech, which were probably unsuitable for Aerospace. I mentioned this to my mom at breakfast and she said, "Then let's go shopping!"

Off we went. I tried on all kinds of trendy, professional-looking office clothes. Standing in front of a mirror modeling a well-coordinated suit, I noticed my long straight hair.

"Mom, I can't go in looking like this," I said. "I haven't changed my hair since I was twelve years old."

"Well, it's so long you can style it in a lot of different ways," she replied.

When we got home I went into the bathroom and put on one of my spiffy new outfits with a pair of pumps, instead of real shoes. I felt like I was walking around on tiptoes. *How do people get used to these things?* I thought. Then I played around with putting up my hair with a few bobby pins and clips. I could achieve several professional looks. Finally, I was happy with the swept-up do I had created and went to the kitchen to show my mom. She loved it.

"Wow, Stacey, that's wonderful, perfect for your new job. With this new look you'll meet some nice, stable engineer and settle down, instead of dating all those flaky musicians."

"Oh, Mom." I sighed and walked down the hall. I was so uncomfortable in those confining clothes with my hair pinned every which way.

As I stepped back into my room, Wesley saw me, screamed, and ballooned out into a full threat display.

"What? What? Wesley, what are you doing?" I said.

Then he attacked. I leapt backward and ducked. He feinted away from me, flew across the room, and landed. He stared at me, gyrating his head around and around, and forward and back. In times of extreme emotion he had a double stomp with one foot: *wackWACK, wackWACK.* He let out a long hissing scream and clacked his beak. Then I saw that he wasn't looking at my face at all, but staring at the top of my head.

"Oh! My hairdo!" I realized he was trying to kill my hairdo. It must have looked like a fluffy blond predator was attached to my head.

I tore out the bobby pins as fast as I could and pulled down my hair.

But Wesley still wasn't convinced. He shook his head over and over again and stared at me with his piercing black eyes. I went over to pick him up, but he dodged me. He was still focused on the top of my head and he screamed again. "It's all right, Wesley. It's all right," I said softly, standing very still. "I'm okay." He flew around the room several times and landed on his perch. He stayed there for the rest of the evening instead of playing, keeping one eye on my head at all times.

"Wesley," I sighed, "when I brought you into my life, I didn't know that I would never be able to change my hairdo." But in the wild, his mate would not have suddenly appeared with a fantastic new feather-do. Besides, I didn't really want to change my hair. I guess the Aerospace Corporation would just have to accept me as I was.

Wesley was nonchalant about some rather big changes, like the move to Huntington Beach. But he was extremely aware of anything on or near me. Later, I had almost exactly the same experience with him the day I went into the bedroom wearing sunglasses, and again when I forgot I was wearing a baseball cap.

Having spent a day sorting out my wardrobe, I still had the rest of the week to explore my new surroundings. Late one afternoon my mom offered to take me on a drive up the Pacific Coast Highway. We had only gone a mile or two when I saw a sign saying "Bolsa Chica Ecological Reserve Wetlands."

"What's that?" I asked.

"It's a wildlife area with walking and biking trails. There was a long, protracted battle to save it from development. It's become a crucial habitat for many endangered species of birds. Most of the other estuaries have been filled in and paved over."

"Let's stop and check it out," I said.

We pulled into a small parking lot. A sign listed all kinds of birds that frequented the preserve. Out in the estuary, the first ones I noticed were the herons and egrets—magnificent waterbirds, slender, graceful, and tall. The great blue herons, at up to fifty-two inches, were almost as tall as I was at five feet, and the great egrets were a bit shorter. Some of them stood peacefully asleep with one foot pulled up, and some waded in the shallows with long fluid strides. Occasionally one would halt and stay motionless for a few seconds and then spear the water, coming up with a flashing silvery prize wriggling in its beak.

Photographers were lined up along one of the bridges, taking evening shots with the pink sky slowly turning into a blazing red and reflecting on the water. It was spectacular. "I'm coming here as often as I can," I told my mom. "I could ride my bike, it's so close."

When I got home, Wesley chirped a greeting. "I saw thousands of birds today," I told him. "They hunt like you but otherwise they are totally different. They're waterbirds, perfectly designed for life beside the sea."

The next day, while I was organizing things in my room, Wesley began to beg at the door, asking to go out. He had a specific sound for this, which he uttered while staring intently at the door crack. Wesley had always begged this way at Wendy's house and would run straight to the bathroom after I let him out.

"Wesley, you don't even know where the bathroom is," I

At about one year old, Wesley begs to go out the door. *Stacey O'Brien.*

said and continued to organize things in the room. Wesley then stood in front of the door and pulled up one foot to wait for me, with an air of resignation. "Okay," I sighed, "I'll open it." I wondered where he would go.

Wesley charged down the hall in his awkward galumph, and without hesitation ran straight into the bathroom. I was amazed. How had he known? Ah. He had heard me go in there to wash my hands and run the water and already knew the location of the bathroom. Owls map their world by sound. I kept forgetting that.

Wesley could hear things that I couldn't even perceive, and sometimes he would freak me out by staring at some spot on the wall and hissing, even going into a threat display. When I went to the spot and pressed my ear against it, I might hear the tiniest sound—perhaps a mouse or bug walking along or chewing. Once, he started going nuts about the trash can, staring at it, doing exaggerated head gyrations, and trying to get off his leash to investigate. I went through the trash piece by piece until I found a small bug walking on a piece of paper, which must have sounded loud to Wesley.

Another time, Wesley attacked a wall with all his might, racing across the room and thrusting his talons into it, then sliding to the ground, making his victory cry. As he landed, he hunched over his "kill," mantling as he would have over a live mouse he had hunted. But I couldn't see anything there. I searched all over—the wall, the floor, the area beneath his feet. He hissed and continued to behave as if he had killed something. Finally I lifted his feet, one by one, and found, pressed into the pink "palm" of the bottom of one foot, a smashed spider. I started laughing as I peeled off the dead spider, "Is this your prey? Are you looking for this, Wesley?" "De DEEP DEEP DEEP DEEP deeple deep!" he responded happily. He had heard that spider walking on the wall and had attacked. I never stopped being amazed and surprised by his auditory abilities.

WESLEY EXPLORED THE bathroom thoroughly. At Wendy's house, the shower had doors that enclosed it all the way to the ceiling, but here we had a bathtub with a shower curtain. He took an immediate interest in the curtain, going under it and behind it, like a kid playing hide-and-seek. He decided that the small space between the tub and curtain would make a great nesting nook and began his earsplitting mating call. I remembered my assignment from Dr. Penfield but realized that in the context of normal life, the task was, well, odd. I decided to wait until my mom wasn't home.

Wesley continued his exploration of the bathroom. He checked the top of the toilet for magazines. Yep, they were there. He seemed pleased. Then he went behind the toilet, which had been one of his favorite spots in the other house. He twittered contentedly. Then he flew up to the counter to check the sink. It looked pretty much the same, so he started his "Please put water in the sink" sound and I turned the faucet on. He swished his face and had a drink.

I went through the bathroom and generally owl-proofed it. At Wendy's house the toilet lid was textured, but here it was slick. Wesley flew to it, slid across and fell over the other side with an outraged squawk. "You're hopeless," I said, picking him back up. I placed a heavy towel over the lid so he could land there. "Owls are not adapted to toilets, Wesley," I told him, carrying him back to the bedroom.

Eager to return to Bolsa Chica, I rode my bike there that afternoon. A small flock of ducks was swimming in the estuary, occasionally dipping their bodies under the water. The ducks were so buoyant that they had to paddle hard to stay down for a few seconds before bobbing back up. Water droplets spilled off their outer feathers like jewels. The air filled with California least terns, wheeling around, calling to each other, and glancing at the water. Every few seconds one of them would dive underneath the surface and emerge with a little fish, shake itself, and dart back into the sky. Although I was nervous about my new job, I could tell that spending time at Bolsa Chica would be the perfect remedy. Near sunset, I rode my bike home along wide stretches of nearly empty beach, where large groups of seagulls were clustered together for the night.

Wesley was waiting anxiously for my return and when I entered the room he immediately begged to be let out again. "Okay, Wes," I said. "I think the bathroom is safe for you now." I let him out and he made a beeline for it. I followed, closed the door behind him, and went to the kitchen to have dinner with Mom. Just as we started to eat, I heard Wesley make a big exclamation of joy. My mom said, "Your owl must be having a great time in the bathroom with all that racket . . ." I dropped my fork and ran down the hall.

When I opened the door, I could hardly believe my eyes. I had accidentally left the toilet lid open, and Wesley had jumped in. He was soaked to the skin, with little wet punk rock feather spikes sticking out everywhere. He looked up at me happily with one wing slung casually over the seat. He had dragged wads of toilet paper in there with him, and it was all over his body in soggy little clumps. Water and bits of tissue covered the entire bathroom. He was shivering violently.

"Wesley! Now what have you done?" I said. He chirruped. I lifted him out of the toilet, heavy with water, and he fought to get out of my arms, kicking and squirming to jump back in the bowl. Now I was soaked and covered with soggy toilet paper. He struggled until he saw me close the lid.

I wrapped him in towels and tried to warm him. "You should have been born a heron, Wes!" He still shivered. This was unbelievable. "There is no such thing as a water owl, you crazy bird." I could not get him dry with towels and it was plain that he was dangerously cold. Dr. Penfield had said that being chilled might kill an owl. I could only think of one way to get him dry, but it wasn't going to be easy.

"Wesley, I'm going to blow you dry, okay? You are cold and you need to get warm, so I'm going to dry you off." Even if he didn't understand my words, my tone of voice and my explaining meant to him that I was preparing to do something deliberate. It seemed to calm him in new situations.

I rarely used the blow-dryer myself. When I turned it on, he tried to fly away from the sound and discovered that his wet wings wouldn't lift his heavy, waterlogged body. This scared him even more and he started running. I caught him and set him on a towel on my lap. I turned the dryer down to low. Every time the warm air hit him he started kicking and struggling. He was so delicate that I let him go when he resisted too strenuously, so that he wouldn't hurt himself. Maybe showing him would help, since he always wanted to do what I did.

So I started blowing my hair. He watched solemnly from the floor, trembling and looking so rumpled and bedraggled I felt sorry for him. I waved the blow-dryer at myself, then aimed it at him on the floor. He startled when he felt the breeze, jumping a little and raising his wings. But then I turned it away from him and back on to myself. "Mmmmm, this feels good," I said. I pretended to groom myself and he began to preen his feathers. I aimed it back at him. This time he didn't jump. He stood still with a look of martyred endurance on his face. I turned it back to myself before he could get scared again. By doing this over and over, I was able to convince him to stand still while I aimed a low warm flow of air at him and dried his feathers. In fact, I thought he looked like he was beginning to enjoy it. *I hope I never have to do that again,* I thought as I put the blow-dryer away.

When I emerged from the bathroom, dinner was cold, of course. I was wet and had bits of soggy toilet paper all over me. Mom dropped her jaw, and I said, "My owl thinks he's a duck."

That night I assembled the recording equipment Dr. Penfield had loaned me. Mom would be out tomorrow afternoon, and I wanted to get some audiotaping done before my job started.

Early the next morning, I rode down to the beach again. I was thrilled to see a large flock of endangered California brown pelicans approaching like a formation of B-52 bombers, zooming over the ocean only inches from the surface with their heads tucked back and their wings fully extended. Spotting a school of fish, they flew up, flipped over, folded their wings tight to their bodies, and dove straight down, like the owl I had seen doing aerobatics five years ago. But instead of pulling up at the last second, they crashed into the water at 40 to 45 miles per hour. Pelicans have small bubbles embedded in their skin that inflate as they hit the surface, like tiny airbags. These protect their body and organs from the impact of diving, as if they're encased in bubble wrap. The fish, however, are not so lucky. A shock wave from a hit at that speed stuns all of them within six feet. The pelican then swims around under water and scoops them up. It's elegant.

When I got back home Wesley was begging again to go to his beloved water park, the bathroom. "Just a minute, Wesley," I said, trying to stall. But Wesley could not be put off for long. In the wild, mated owls don't stop each other from doing what they want. So once Wesley set his mind on something, there would be no peace until I gave in. I quickly carried the recording gear into the bathroom, pointing the microphone at the area between the shower curtain and tub.

I untethered Wesley and he rushed down the hall right to his favorite new nesting area. He started his call, and in that small room it was louder than ever, reverberating off the hard bathroom surfaces. I clicked on the recorder and sat by quietly while Wesley repeated his harsh, relentless cries. I was able to record a good long chunk of his mating call before he suddenly jumped on my arm. The cry turned into the squawking parrot sound he always made and at last the one great screech at the end. Perfect.

I was still a little confused as to why Dr. Penfield wanted this recording. Earlier in his career, he had put an infrared camera inside a barn owl nesting box and recorded their sounds for two years. Surely he had heard every possible vocalization an owl could make by now. But I was happy to have succeeded in my assignment.

I gathered up the equipment and opened the bathroom door. There stood Mom.

"What on earth is going on?" she asked.

"Mom!" I yelped. "Uh, that was Wesley's mating cry."

"Well, for goodness' sake, he sounds absolutely hysterical. The poor creature, there isn't a female owl anywhere around here for him."

"Mom, Wesley thinks I am his mate. He's making those calls to me." I could see her processing that statement.

"I hope you don't tell people about this, Stacey, because they wouldn't understand," she said, as she turned on her heel and went to her bedroom.

As I prepared to get into the tub that night, I tethered Wesley in my room. "I won't be long, Wes," I said. "I'm going for a bath." But it was not to be. When he heard me in there splashing water, he started begging insistently, making his series of owl comments, exclamations, and lectures that let me know just how put out he was that I was leaving him behind while I had fun. I couldn't stand it anymore. Perhaps if I just let him into the bathroom, he would play with a magazine while I bathed. Dripping, I climbed out and looked down the hall, the coast was clear. I ran into the room, released Wesley, and raced back into the bathroom with him galumphing joyfully behind me, wings out like an airplane as always. Then I closed the door, handed Wes a magazine, and slipped behind the shower curtain.

I sat down in the tub to resume my bath when suddenly, by the side of my head, there was a *wump* into the shower curtain. Then another *wump . . . wump . . . wump*. Wesley was flying right into the curtain, trying to find a way in. Then he flew against it, gained purchase with his talons, and flapped his wings hard. He reverted to the old tree climbing antics of fledgling owls and powered right on up. His eager face appeared at the top of the curtain. *Uh oh*, I thought. He looked down at me and his surprise was classic: head gyrating, wings lifted, he fairly jumped out of his feathers to see me sitting in a *full tub of water*. It must have looked like an ocean to him. He leaned far, far over, holding tightly to the curtain rod. "Not for owls!" I cried. "Not for owls!"

Wesley dived straight down. As soon as he felt the hot water he started flailing. In one motion I swooshed my hand under him and lifted him out, shutting the curtain quickly. *Well, he won't do that again*, I thought. But before I knew it he was up on the curtain rod and flinging himself back into the water. This time he extended his talons just before he hit. "Wesley! Stop!" I said, scooping him out again before he had a chance to sink. "You're not a pelican."

Owls just don't throw themselves into the water like this. It's completely unheard of. Barn owls are not waterproof or adapted for water and normally avoid it altogether. They don't even drink it, because they get all the fluids they need from their prey. Wesley's antics would mean certain death in the wild.

I quickly finished my bath, drained the tub, and let him explore it. Wesley began complaining that the water was gone with tiny hisses under his breath. Then he found the dripping faucet and stood under it, begging.

"Okay, Wes, just a little bit."

I turned on the faucet. He wanted to dive right under the spout but I held him back while the tub filled up to about an inch. I turned Wesley loose and he kicked and splashed and played—there's no other word for it—joyfully. From then on, this was his favorite place to hang out, with or without water. He liked the cool tub floor on his feet on hot summer days. Sometimes he would stand in the shallow water, pull a foot up, and sleep! Other times he just waded around, taking little face baths and sipping water whenever he felt like it.

When I told Dr. Penfield about all this he was amazed. "This has never been observed in barn owls," he said. And at that time, it hadn't been. Bernd Heinrich has written about a great horned owl that he observed taking a bath in a pond behind his cabin, but Wesley's habits were confounding.

Now I had a real dilemma—how to bathe without Wesley's constant interference. In any other situation I could tether him, but he had become obsessed with water. He went crazy if I attempted to bathe without him. I tried bringing him in and putting a doggie bowl down, but to no avail. I tried leaving the shower curtain open and putting a towel over the edge for him to stand on, which made it easier to keep him out since he wasn't diving recklessly, but he was still trying to jump in from the side. Very unhappy with my selfishness, Wesley did a hiss-scream of protest when I caught him and returned him to his sitting spot over and over again.

I thought I'd try taking a shower instead. Maybe he wouldn't be quite so obsessed if there wasn't an actual pool to jump into. I was wrong. Wesley climbed to the top of the curtain and started diving randomly, talons spread and facing down, flailing around for a place to land, a frantic look on his face. The only possible landing spot would be somewhere on my unprotected skin and that would be disastrous, so I dodged and caught him in midair. What was I going to do in the shower with an owl in both hands? I pushed him outside again and tried to hurry up and finish, but he kept reappearing, his eager face all excited about this new experience of being hit with streams of water. I started waving my hands above my head, feinting this way and that as he tried to find an opening through which to dive. Aha. This was working. One hand was waving erratically over my head so he couldn't plot a diving trajectory, the other trying to soap up and shave my legs and get clean. Sometimes I had to stop and wave both hands if he started looking like he was going to just go for it no matter what. That usually made him back off a little and I could go back to dancing with just one hand waving over my head. I felt absolutely ridiculous, thankful that no one would ever know about this.

One day, during this typical scene, I accidentally flicked water at him and he got a look of "Eureka!" on his face. He appeared to ponder this for a moment, then leaned way over with his wings out. Because of the incredible strength in his legs and talons, he could hold himself on the curtain rod at an extreme angle. He was almost upside down. "Oh, no, you don't, Wesley. Don't dive," I said. But he wasn't trying to dive. He spread out his wings, then lifted each feather up off of his body and rocked gently back and forth. He wasn't threatening me. Was it the flick of water that he wanted? I filled my hands with water and tossed it up over him. He almost swooned with joy. That was it! He was asking me to "water" him like a plant. "That does it, Wesley, I'm taking you to a counselor. Owls are not waterbirds," I told him. Apparently he didn't care, he just wanted a shower like I was having and he'd finally figured out how to get one.

I was still trying my best to avoid getting him soaked. I could get through much of my showering if I would just flick water at his fluffed feathers as he held this crazy position. Then he'd poof up, shake, preen, and go back to his new showering pose.

The next night I decided to fill the tub up to his hips. Then he did the most surprising thing of all. First, he took a regular face bath, dipping his face into the water then shaking his head fast, sending a spray up around him. Then he began to bend his knees slowly until he was up to his chin. I watched, fascinated. He splayed his feet apart on the slippery bathtub floor, immersing himself further. Then he opened his huge golden wings as wide as he could and let them sink into the water. What on earth? Suddenly he was a flurry of action, shaking his entire body from head to toe, face in the water, wings flapping, sending a huge spray across the wall and over the side and even onto the ceiling. I was now getting a shower while he was taking a bath.

Wesley looked up at me as if he had just noticed I was there. He had that same self-satisfied expression I'd seen when I had found him in the toilet bowl. He dipped, shook, and flapped again, sending much of the bathwater into the air and over the

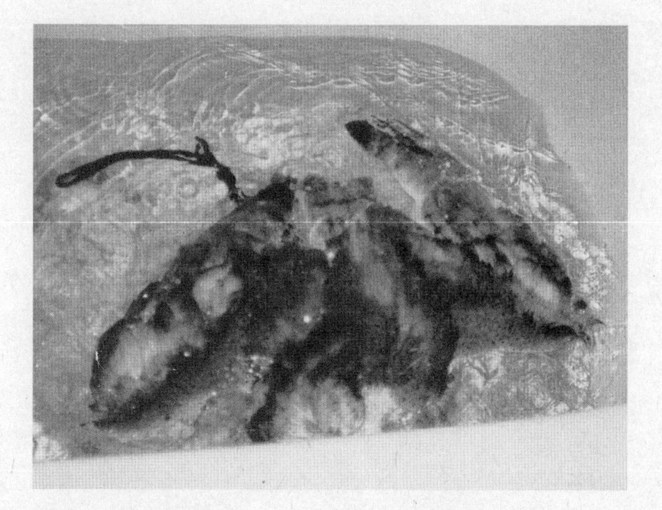

A very rare behavior for a barn owl, Wesley has immersed himself completely in the water. *Stacey O'Brien.*

Standing peacefully in the tub. *Stacey O'Brien.*

After a bath, Wesley admires himself in the mirror. *Stacey O'Brien.*

side. What a show. At the end of his bath, I had to intervene and pull him out, to his loud protestations. He twisted and struggled to get back in, and I had to drain the bathtub before he would calm down. He was just like a little kid who wants to keep playing in the water even though his lips have turned blue.

Once the tub was drained, I lifted him up to the counter and Wesley admired his soggy self in the mirror—wings out, turning to different angles to see how he looked. He chirruped happily at his reflection over and over. He couldn't take his eyes off himself, holding one pose for a while, then shifting and holding the new pose, like a guy in a body-building contest. Then he started shivering. "You're the only owl in the world that needs a blow-dryer," I said, preparing myself for the long process of convincing him to stand still.

But this time when I turned the blow-dryer toward him,

Wesley had another one of his "Eureka!" moments. As I stood still, pointing the stream of warm air at him, Wesley bent forward and fluffed his feathers up off his skin just as he would do while standing on top of the shower rod. Then he moved from side to side and around, adjusting his body so that the air hit him all over and in different places, fluffing up the feathers here, then there, then back again.

Soon he was begging for baths every day. I would fill the tub to his hips, he would dive in, talons outstretched, then do exactly what he had done that first time, spreading his wings and wiggling his entire body in the water just like a tiny bird would do in a birdbath. Except he was a big owl. He always made a mess, soaking the entire bathroom.

"Wes," I finally told him, "you may not be adapted for water, but in your heart you really are a waterbird, as much as any bird at Bolsa Chica."

At about eight years old, Wesley dives feet first into the bathtub.
Stacey O'Brien.

12

Deep Bonds

ONE HOT SUMMER night, I was lying on the bed next to the window, which I had opened so I could enjoy the ocean breeze and listen to the waves. As always, Wesley was chattering and exclaiming to me about all kinds of things as he played on his perch. Outside, I heard a screech, very soft and very, very close. I sat up in the dark to come face-to-face with a lone female barn owl hovering right outside the window, looking in at Wesley. I couldn't believe my eyes, a wild owl inches from my face! I held my breath as she hovered then flew to a nearby fence to rest, as hovering is hard work for an owl. She continued to offer soft screeches.

Unsure of what was happening, Wesley let off a loud screech in response. She liked what she heard and came back to hover. I remained completely still so she wouldn't be afraid. But she didn't seem too concerned about me; her focus was on Wesley.

She was beautiful. Should I let her in? What if I let her in and then closed the screen again? Would she panic? Would they mate? Then what? I could keep her and try to tame her, but I couldn't imagine submitting a healthy wild owl to a life of captivity. And Wesley had no idea how to survive even fifteen minutes outside. He was afraid of the trees when they moved, as if they were monsters. Even if I had started rehabbing him the way they do with owlets that have been fed with owl puppets, he still would have been imprinted on humans and extremely vulnerable to them. He would have come down to humans when it was not appropriate to do so, assuming that they were all friendly, and probably would have been shot as a trophy by some ignorant person. He had not had the opportunity to learn from owl parents what to avoid—traffic, power lines, great horned owls, fire, so many dangers.

In rehab centers they teach the owls to hunt by introducing dead mice first, then pulling them on a string to get the young owl to chase them, then eventually using living mice. Also, the owl has to be exercised to build up enough muscle to have the endurance to hunt for hours at a time. Even with birds just in from the wild for a short recuperation, it's essential to get their muscles and endurance back into wild condition. In rehab centers where I had worked, we would chase the birds while waving towels over our heads to keep them flying in order to gain strength. We were exhausted, too, after these training sessions.

I'd have to do all that with Wesley before I could even think of letting him go in and out, and I'd have to move to an area far from traffic and dangers. I already knew that Wesley could never hunt enough to provide for a nest. So he'd always have some dependence upon me, and if he were flying outside, I wouldn't be able to keep him safe. No, it wasn't possible for him to go off with her, and even if I could have kept his female admirer and tamed her, then what would I do with the babies when they grew up? It would be a real mess eventually. Yet in that moment I felt an ache in my heart—I wanted them to be together. I wanted to see them together. She seemed to be from some other world, like a fairyland—a visitor from the other side.

I decided not to interfere. I was content to enjoy the magic as the two owls talked to each other. Over and over she returned to the window. Finally, she flew off and I lay awake in the dark feeling more alive than I'd felt in a long time, blessed to have been so close to this beautiful wild creature. This was the first but not the last lonely little lovesick female to visit, attracted by Wesley's calls.

MY JOB AT Aerospace turned out to be perfect for me, and I didn't end up needing those fancy clothes I had bought after all. Sometimes my work entailed climbing around under the floor of the computer room laying cables. In fact, my office mate wore the same thing every day, a pair of New Balance sneakers and a USC sweatsuit. It was a wacky, think-tank type of atmosphere much like Caltech. I loved it.

I was also enjoying life at my mom's house. Her dancing partner and boyfriend, Wally, came over to visit one night and as I went downstairs to defrost Wesley's mice for the evening I saw that my mom and Wally were in the kitchen about to prepare dinner. Wally didn't know about Wesley. I took Mom aside and whispered, "What should I do? I need to feed Wesley!"

"Just take the mice out, hide them in a paper towel, and stick them in the microwave. He'll never notice," she said.

Okay. I covertly slipped the mice out of the plastic bag and quickly wrapped them in a paper towel. But as I went to put the four mice in the microwave, Wally walked over to the stove, located beneath the microwave, with a large black skillet.

"Oops, 'scuse me," I said, startled. I fumbled my hold on the icy white mice, which slipped out of the paper towel into the skillet with four loud plunks. My mother and I froze. Wally just stood there, skillet in hand, staring at the mice with an uncomprehending look. Then both Mom and I came to life, shot out our hands, grabbed the mice, all the while babbling things like "Okay! Hey, Wally, what kind of salad dressing do you want? What do you want to drink? Why don't you get yourself a glass," hoping to distract him. It worked.

Just recently my mother and I confessed to Wally about the mice-in-skillet incident and he replied, "No. I don't believe you guys. I was looking right at them? I don't remember a thing!"

NOW THAT I was making better money, I could go out more often. I went out to eat one night with a large group of friends and found myself sitting across from a man named Guy Ritter, a drummer and the lead singer in a hard-core heavy metal band called Tourniquet. Guy and I hit it off right away and he asked me on a date. I had finally found a man who could scream louder than Wesley.

The first time my mother heard a Tourniquet CD, with Guy's low growls and high screams, she sat down as if weak in the knees, saying, "Oh, Lordy, Lordy." When I told her we were dating, she brought her hand to her forehead and said, "Another musician!"

Guy and I had a lot of fun going to his heavy metal gigs with mosh pits, where adolescent boys flung themselves into each other, butting chests, banging their heads up and down. Guy would head-bang so much during his concerts that the next day he'd lie on the couch with ice packs around his neck, moaning.

I didn't tell Guy about Wesley. He thought I kept my door closed religiously because my room was messy. I don't know what he thought about the strange noises emanating from there, but he had the good manners not to ask.

My mom had developed a good relationship with Wesley by standing in the doorway and talking to him in a sweet voice. He threatened her if she tried to come in past the door, but he still enjoyed having her stand just outside his territorial boundary, talking to him. He would respond with chirps and chatter, relaxed enough to stand on one leg, groom himself, and eat in her presence. She began to offer to take care of him so that I could go off on trips. She had to wear eye goggles, a heavy coat, gloves, and a helmet when she went into the room, so those were by the door in case Wesley ever got off his leash and she had to chase him down, catch him, and put him into his carrier. Thankfully, this never happened, but she was prepared.

Mom could defrost his mice and, using barbeque tongs, toss each mouse to the platform on his perch, where he'd pounce on them. While he was distracted with his new meal, she'd use the tongs to collect the old scraps from yesterday's meal and any mice that he'd thrown off the perch, put them in a garbage bag, and take them out of the room. She had quite a nice system in place, I thought. And it gave me a new measure of freedom because not only did I trust Mom, Wesley did, too, and seemed content with her company when I was gone. She had become part of his "inner circle"—the people he would trust to a certain level and who could even touch him if I were holding him and they approached quietly and gently. I could feel Wesley trembling, yet he seemed determined to allow them to pet him no matter how frightened he was, as if he were actively and deliberately fighting his instincts.

Guy and I went camping together and took road trips, laughing and joking, playing loud music and singing. I told him, "You have to meet my grandparents. My grandfather has been a drummer all his life. He played in big bands during the Depression when he was only thirteen years old. He made enough money to support two families, his own and my grandmother's, saving both families from losing everything during the Depression—and he was just a kid himself. Then, in the '40s and '50s, Grandpa played in big bands with legendary performers like Frankie Carle and Horace Heidt. He raised his sons to be drummers. So there are a lot of drummers in my family."

We went to visit my grandparents and they liked Guy a lot. After making him feel welcome, Grandma went into the kitchen to make tea, and Grandpa and Guy started talking about music and telling jokes. I sat gazing at Grandma's collection of hundreds of owl figurines. Having spent many years on the road with Grandpa, she'd buy owls from local artisans wherever they went. Friends and family also had been giving her owl memorabilia for years. I had always enjoyed visiting Grandma's house, and even as a child, I was fascinated by her owl collection. But now, of course, it held special significance for me. Because of my fear of those anti-captivity animal extremists, I never mentioned Wesley to anyone, and nobody in the family but my mom and sister now knew I was taking care of a live owl. The thought of losing him was more than I could bear. As much as I wanted to tell Grandma, I held it in, worried that she might accidentally reveal his existence.

When my grandmother came back into the room, Grandpa's eyes lit up. He said, "Here comes my queen!" and looked completely smitten, as if just now discovering her great beauty. They never grew "old" to each other. My grandparents had the best marriage I'd ever seen. When my grandparents were on the road while Grandpa was playing in the big bands of the '40s and '50s, the bands attracted such huge enthusiastic crowds that even after gigs, the performers sometimes had to spend the night backstage because it was impossible to get to their cars.

All that attention and glamour didn't sway my Grandpa one bit. He was absolutely devoted to Grandma and never looked at another woman. Grandpa lived the Way of the Owl.

On the way home, I decided to tell Guy about Wesley. I turned to him and said, "Um . . . I have a secret. I need to know that you won't tell anyone, not even the other band members."

He looked concerned. "Okay . . ."

"I have a barn owl living with me in my bedroom."

"No . . . really?" he said, incredulous. "You have an owl?"

"Yep. That's what all those weird sounds are coming from my room."

"That's an owl? I thought owls hooted."

"Not this one. Barn owls don't hoot."

"Well, what are we waiting for? I can't believe you've never showed him to me. I want to meet him. Come on, hurry up, hurry up!" He floored it all the way home.

"Okay, Guy, now calm down. Wesley's a wild animal and he's not used to men. You can't breach the doorway because that's his territory and he'll go into attack mode."

"He . . . he attacks people? Has he ever?" he asked.

"Yes . . . uh, well, he attacked my last boyfriend, actually."

"Great," he said. "Well, tell me what to do."

I took a breath, "You should tell him he's handsome."

Guy looked at me, "You're kidding, right? You want me to tell him he's handsome?"

I coached him on owl etiquette just outside my bedroom so that Wesley would know we were there and understand that I was relating to this other person. Then I slowly opened the door.

"Hi, Wesley! Someone's here to meet you."

Wes had heard Guy's voice on and off for months so he

knew there was another person in my life. They looked into each other's eyes.

"Hi, there. You're so handsome, Wesley! You're beautiful . . . My gosh, Stacey," Guy said in a hushed voice, "he's gorgeous. I've never seen a golden and white owl before. This is an owl? Wow, he *is* actually handsome, to tell the truth. He hardly looks real."

Many people were surprised by Wesley's looks because barn owls are very different from all other owls and people's expectations of what they look like. Wesley did not have ear tufts and he was not darkly colored like most other owl species. Most owls have colored eyes with black pupils, but Wesley's eyes were solid obsidian black. His face was exotic, heart-shaped and pure white. The contrast was stunning and caught most people off guard.

To my amazement, Wesley did not go into his attack no-nos but just watched Guy carefully. Guy stood at the door for a full hour talking gently to Wesley, and they seemed to understand each other. Wesley had never responded this well to anyone other than Wendy and my mom and sister. I was very excited that things were turning out so well. Guy thought Wesley was the coolest creature on earth and I could see that he would have no problem living with him, if we got serious and decided to marry. (Women generally look at their boyfriends as potential husbands and fathers, no matter what we say.)

After Guy and I had been going together for several months we decided it was time for me to meet his family in Oregon. Guy's dad, although not actually a logger, worked in the logging industry, which made me nervous. He was an engineering specialist in the factory equipment that turned logs into paper, so logging provided him with his livelihood. There was a well-publicized antagonism between biologists and loggers at the time. Biologists were warning the public that the old-growth forests, a delicate habitat that can't be replaced, were disappearing at an alarming rate. The streams and rivers were silting and warming up, destroying the salmon runs and the entire ecosystem because of runoff from clear-cut areas. The apex predator of these forests, the northern spotted owl, was endangered. (The apex predator is the very top predator in the food chain of an ecosystem. In the Australian bilabongs, for example, it's the crocodile.) When the apex predator is thriving, then so is the environment. But when the predator is faltering, biologists know that means the entire system is falling apart.

Most of the loggers didn't understand this "canary in the coal mine" connection and thought the entire issue was about saving the owls, rather than their habitat. Because the loggers had been told to stop destroying ancient forests before the forests were completely gone, they would lose their livelihoods sooner than if they kept cutting down trees until the entire ecosystem went extinct. Focusing on only their own livelihoods, they didn't want to be told what to do, got angry, and took it out on the owls. Some loggers actually killed them and tied their bodies to the backs of their trucks to protest the government's and the conservation groups' efforts to save this ecosystem before it was destroyed entirely. They didn't understand—or they just chose not to—and they reminded me of the buffalo hunters of the nineteenth century determined to hunt down every last animal. They failed to see that they were going to have to find something else to do anyway after the last buffalo was gone.

As soon as we drove into Oregon, I began to see threadbare tree lines with nothing behind them but stripped land. Hundreds of miles of priceless virgin forest had been wiped away, with just a thin veneer of trees left standing along the roads to fool tourists and others into thinking the forest was still intact.

It wasn't. Oregon would have been one of the most magnificent and sought-after tourist spots in the world if the old-growth forests had been saved.

Guy and I decided we just wouldn't mention my owl to anyone. The people in his community wouldn't understand. His own parents might not understand. Guy, on the other hand, had spent his youth hiking and camping in the old-growth forests. He loved them passionately and did not want to see them destroyed.

By the time we got to his parents' house at around 4:00 a.m., we were exhausted from driving straight through the night. We crept in silently, and I went to the guest room while Guy went downstairs into his boyhood bedroom. He left a note for his parents saying when we'd arrived and please not to wake us up. As nervous as I was about making a good impression on his parents, I still had no trouble drifting off to sleep.

I woke up when Guy's mother, Aileen, poked her head into my room.

"Stacey? Hi. I'm Aileen, Guy's mom. I'm so sorry to wake you up like this, especially since we haven't been properly introduced, but your mom is on the phone."

I was wide-awake now. *Wesley! Was he okay?*

"She says she, uh . . ." Aileen looked at a note in her hand. "Well, I'm sure I misunderstood her, but I think . . . she's asking me to ask you what . . . uh . . . what setting to use on the, well, the . . . the microwave. The microwave?"

I nodded encouragingly.

"On the microwave for the . . . I couldn't understand her very well, but I think she said mice. Mice?"

"Yeah," I mumbled.

"Okay then. Um, what setting to use to defrost the mice? Defrost?"

I nodded. So much for owl secrecy.

"Okay then. What setting and for how long should she . . . uh . . . well . . . uh . . . defrost them . . . for the . . . *owl?!*" she blurted out.

"It's for Wesley. I wrote all this down for her. She was only supposed to call for emergencies. I'm so sorry."

"Oh, well, that's okay, but what should I tell her? She's on the phone right now."

Great.

"I'm sorry, Aileen. She's supposed to set the microwave to defrost for twenty seconds. Then she's supposed to take the baggie with the four mice in it out and let it sit until there are no hot spots and no frozen spots. She can kind of mush the baggie around in her hand to make sure they're completely soft so he has no trouble swallowing them whole."

She looked like she was going to faint.

When she left to deliver the message to my mom, I buried my face in my pillow.

"Nooooooooo," I groaned.

It turned out that owl secrecy wasn't necessary. Guy's parents were wonderful. We had a lovely ten-day visit way out in the country next to the McKenzie River, hiking around the old-growth forest with long flowing curtains of moss hanging down from the trees and huge yellow slugs and tiger salamanders. We picked wild berries and drank from ice-cold streams. It was a kind of paradise.

Aileen became so interested in Wesley that she decided that if I was recording his vocalizations at home (I left out the part about him mating with my arm), I should also have recordings of the now very endangered spotted owl, which they occasionally heard when they were sitting out on their deck. She and Guy were sure they could get the sound on a tape recorder if they tried. I thought the folks at Caltech would be thrilled to get a recording of the vocalizations of this species.

I went to bed fairly early, and the next morning I woke up to find a tape on my nightstand labeled, "Sounds of the Northern Spotted Owl."

I got ready as quickly as I could and went to the kitchen, where the family was already gathered for breakfast.

"What's this?" I asked.

"Guy and I went out last night and captured spotted owl sounds for you!" Aileen said.

"Really?" I was amazed.

"Yes, we drove around until we found a good spot and recorded them."

Guy nodded enthusiastically as he stuffed his breakfast into his mouth. I was really touched. Here this family in the logging industry had gone to great lengths to accept and include me.

On the way home, we visited Wendy at her new house, a cabin in the Oregon woods. She was glad to meet Guy, since I'd been telling her all about him on the phone for months.

"Come on, Stacey, wait till you see this!" With Annie at her side, Wendy led us out back to a fenced pen that held Courtney the dog, a gigantic brown Norwegian rabbit, and a young fawn. The fawn was beautiful, tawny with a few spots and large brown eyes. We all took turns bottle-feeding her.

"The kinship between these animals is amazing."

As Wendy said this, the older female bunny, who was the biggest rabbit I'd ever seen, named Fierce Bad Rabbit, pawed at the fawn's knees and the fawn buckled and lay down. Fierce Bad Rabbit then hopped onto the fawn's back and stretched out to take a nap.

"Two weeks ago the fawn got out of the pen and a strange dog spooked her and she ran off. The next day, Courtney disappeared and all we found was her broken collar. Annie and I were heartsick. Three days later, Courtney appeared in the yard with the fawn trotting behind her."

Wendy laughed. "In the wild, a deer and a canine would be enemies. When the fawn is older, I'll take her to a captive refuge. She's a nonindigenous Sika deer. They can't be released because they might interfere with the indigenous species here. But I know Courtney and Fierce Bad Rabbit will miss her."

The fawn got up and grazed on some of the grass at her knees and Courtney began to graze the grass with her, but the dog's lips curled as if to say, "I don't know what that deer sees in this stuff."

Wendy's husband barely spoke to us throughout our visit. I could tell from both Wendy's and Annie's faces that things might not be so good with him. I was sobered by what I saw and worried about my friend.

After two days of driving, Guy and I completed our long journey home. At first I was worried about Wesley's reaction. I had never left him for this long. But my mother had spent a lot of time talking to him so he wouldn't feel lonely and abandoned. He was relatively calm when I got home and had not gotten crazy wild as experts warned he would if I left for too long. Yes, he was a little wilder, a little harder to handle, and a little bit more nervous, but that was all. He tried to act cool and a little distant—he sulked a bit, too, and didn't carry on about greeting me but was so happy to see me that it didn't last long.

Wesley always slept on the side of his perch nearest to me, facing me. But on my first night back from the trip, he slept on the opposite side with his back to me, the way he did when he was angry. I tossed around and couldn't sleep. Finally, at around

3:00 a.m., he moved back to his old spot as close to me as possible, and I fell into a peaceful slumber.

I thanked my mom profusely. How many mothers would put up with feeding mice to a wild animal while her daughter was gone, especially since she had to wear eye goggles and use barbecue tongs to feed him, because he would attack her if she got too close? Not too many.

The next night, Wesley and I were back to our old ways. He did everything he could to lure me into a mating frame of mind but to no avail. He made elaborate nests, with magazines that he selected and pulled off of the top of the toilet, then dragged into the cupboard and shredded apart. He then commenced his incessant, ear-piercing nesting call. Ah! That reminded me of the tape of the spotted owls given me by Guy's mom. I could hardly wait to hear it and I wondered how Wes would react.

I popped the tape into the player and settled back to listen.

"Okay, here, Mom, down here."

"Where are you? *Ouch!*" [*Bushes thrashing.*]

"Turn on the flashlight, Guy."

"No, Mom, I don't want to scare them away."

"Oh, here you are. Okay. Start taping them."

"I am. Shhhhh . . ."

[*Long silence.*]

"Are you getting that?"

"Yeah, this is great . . . she's gonna love this."

[*More silence.*]

[*Crash, shriek, crash, rustle rustle rustle of bushes.*]

"What was that??"

"I don't know, I don't know, just go. Go go go!"

[*More thrashing.*]

"Ouch! You just stepped on me."

"Well, don't stop like that."

"I can't see. That's why I stopped."

"I think that was just a deer."

"Well, it better be just a deer! Get me out of here."

"Don't push me."

"Sorry."

[*More thrashing around in the underbrush and oooches and ouches and finally the ding ding ding of a car door.*]

"Get in! Get in! Just go, go!"

"Okay, okay. I'm going, geez!"

[*Click. End of tape.*]

DURING THIS PERIOD of my life, deep bonds were forged and broken. Wendy's marriage ended, and she and I spent many tearful nights on the phone. Through this difficulty our friendship became stronger than ever. And although Guy and I had wonderful times together, our dating relationship began to wane, which was painful for me. I visited Grandma during this time and told her that Guy might not be the one.

"Stacey," she said, "I knew your grandpa was in love with me when he dipped my braids in the inkwell of his desk. He sat behind me in the fourth grade. We never dated anyone else."

"You and Grandpa are like a pair of owls," I told her. "I have always hoped for a marriage like yours." We had a long visit, and as I gazed again at her owl collection, I thought about telling her my dearest secret.

"Grandma," I said.

"Yes?"

"I have something to . . . to show you. You must come visit me sometime."

"I'd love to," she said.

Guy and I broke up, though we are still friends to this day. That night I cried into Wesley's feathers and poured my thoughts out to him. As always, he watched me with his deep black eyes and listened to every word I said.

I had been at Aerospace for almost five years and had become an expert in UNIX operating systems but was still being paid an entry-level salary even though my skills would bring nearly four times that in private industry. I struggled with whether or not to leave, exploring my options when an event of historic proportion overtook my decision.

The Rodney King police brutality verdict hit the streets the afternoon of April 29, 1992, and Los Angeles soon erupted into rioting not far from where I worked. Black smoke blew ominously past our office windows. I took our computer servers down to preserve data and directed members of our department to organize into caravans to drive people home. We searched for anything useful as weapons for self-defense. Among the staff there was still debate about whether it was safer to stay or go. Then the power shut off.

We quickly evacuated the office and our caravan threaded through out-of-control traffic, with cars running stoplights and blocking lanes, even driving on the sidewalks. From the freeway we watched Los Angeles burning. Almost every block as far as we could see had a building on fire, as if the whole city had been bombed by some foreign air force. The freeway was almost stopped and it was a vulnerable feeling, just sitting in our cars, hoping that the rioters didn't come over the walls and start killing the unarmed people in the cars. No one knew what was going to happen next. There wasn't a police car to be seen all the way home. Huntington Beach was a ghost town.

I was worried and shaken but happy to be safely home with

Mom and Wesley. He was hungry, as usual, so I went to the freezer and discovered I was out of mice. This was not good. I drove around town, but all the businesses were shut down and the nearest pet store was dark, though cars were still in the parking lot. I banged on the glass double doors.

"Jason! Jason! Are you in there? It's me, Stacey! I need mice!" I saw movement in the shadows of the store and Jason appeared and let me slip inside.

"Hi, Stacey, go get them, but be quick. Most of the animals are out, but we're still evacuating. I just got a call from a friend of mine in LA and the rioters burned down his pet store."

I headed for the back of the store where the mice were but stopped short. In every aisle a man was sitting on the floor with a gun across his lap. "These guys are spending the night to guard the store," Jason said.

As it turned out, Jason's animals were under no threat; the rioting never spread this far south. In the days that followed, Los Angeles continued to burn, and for months tensions were high. The commute into LA no longer seemed safe, and the whole incident finalized my decision to leave Aerospace. Wesley and I moved shortly thereafter to a wonderful community in north San Diego County, where the air was clean and crime was low, and there were so many computer-related jobs that I could almost take my pick. I found a two-bedroom apartment in La Costa, up in the hills above a small river gorge. At night I could hear coyotes and wild owls calling from the canyons around my new home.

I was reading a book at dusk when I heard something that made me jump out of my chair. It was the same earsplitting mechanical noise Wesley always made when he wanted me to join him in a nest, but it was off in the distance. Before Wesley was a part of my life, I would never have tuned in to such a sound. I dropped my book and ran outside. In the darkening sky far above me was a lone male barn owl, swooping gracefully in large figure eights, screeching in that unearthly manner all the while. From a nearby tree a female owl screamed in response. Then she shot straight up into the air toward the male. They circled each other closer and closer, and locked talons. They spun together, broke off, circled in the air, locked talons, and spun again. Their magnificent dance continued for several minutes before they flew off together, to mate.

UNCLE WARREN PHONED to tell me that Grandma had been admitted to the hospital. Though it was supposedly nothing serious, the whole family visited her often, myself included. Grandpa was always there early in the morning until very late at night, when they kicked him out. When they'd let him, he sometimes spent the night. But Grandma did not get better. Her kidneys began to fail. On my last visit with her, she was heavily drugged. I bent down and told her, "Grandma, I wanted to surprise you when you came to visit me, but I'll just tell you instead. I have a barn owl. His name is Wesley. He lives with me in my bedroom."

Her eyes lit up for a moment and she said, "So did I."

I didn't know what to think about her comment, and she didn't elaborate, but fell asleep.

She died shortly after that.

A few months after Grandma's funeral, I told Grandpa that I had a real live barn owl named Wesley.

"I tried to tell Grandma about it, but I don't think she understood."

"You have a barn owl?" My grandpa had a funny look on his face.

"Yes, and Grandma said she had one, too, but I think she meant in her collection."

"Oh, no, Stacey, we had a barn owl. A real live one. His name was Weisel."

"Weisel??" It sounded a lot like Wesley, the way he said it.

"Yeah, you know, like wise ol' owl. Wise Ol' Weisel."

"Wow, Grandpa, that sounds an awful lot like Wesley."

He thought for a moment. "You're right, it sure does. It sure does. Well, I'll be."

"How did she end up with Weisel?" I asked.

"She's always been an owl fanatic. Some neighbors found a barn owl being savaged by two dogs and ran between them to rescue the owl. The dogs bit them up pretty bad, too, but the neighbors finally got the dogs off. Then they didn't know what to do with the owl."

"Did the owl attack them?" I asked.

"No, he seemed to understand that they had saved his life. I think he was in shock pretty much. Anyway, one of the neighbors remembered that your grandma collected owls and thought she might take this guy in. They were right of course. His wing was really a mess so we rushed him right to the vet. In those days there was no wildlife rehab place so it was up to whoever found an animal to make sure it was taken care of properly. The vet fixed him up as best he could. The wing wasn't broken and Weisel could still fly, just not very well. Then the vet sat us down and told us everything that had to be done to care for the owl, and he told us it would never survive in the wild again so it was up to us."

I shook my head in wonder that my grandmother had had an unreleasable barn owl, too.

"Your uncle Warren and I built him a huge aviary around a tree!" Grandpa continued. "That guy had it pretty good."

"What did she feed him?" I asked.

"Mice! What else do you feed an owl! And I fed him, not your grandma. She wasn't too keen on those mice." He smiled.

"So really Weisel was your owl as much as hers, wasn't he?"

"Yep. He sure was. He had a grand old time in that big tree of his. We also put in extra perches for him. The aviary was huge. We had him for the rest of his life, old Weisel."

"Grandpa, why have I never heard a word about him until now?"

Grandpa reached over and hugged me.

"Stacey, all of your grandma's animals were like human children to her. Whenever one of them died, she was so devastated that she never spoke of them again."

At home I told Wesley of my sorrow as I had done many times in our years together. "Wes, I am so sad that my grandmother and I never figured out that we both adopted barn owls." He preened himself and then my face for a moment with gentle movements of his beak.

Wesley groomed on and off all day, but about twice a day he had a serious total body grooming session during which it was hard to distract him. He went over every feather and it took at least an hour. I plunged my face into his sweet-smelling feathers while he did this, and followed his beak with my nose.

His body was perfectly synchronized. When Wesley pulled a feather that was ready to come out, he invariably also pulled out the exact opposite corresponding feather on the other side. If it was the third secondary flight feather on the left side, a few minutes later out would come the third secondary flight feather on the right. And Wesley could move entire sections of feathers into reach by shifting the skin underneath. One whole side of his body would suddenly move closer to the front so he could groom it with his beak. Then that patch of feathers would ripple back to where it belonged. I knew his grooming routine intimately.

"Someday I will tell Grandma all about you, Wesley. Perhaps we will walk together, Grandma and Weisel, you and I."

Wesley opened his wings and began to arrange his long, breathtaking flight feathers. These were not replaced very often—only one pair of feathers every six weeks or so. When the time came he would work at loosening one of these beauties—pulling his wing way out and yanking and worrying the feather until it would finally come free. Then he would hold it and play with it. I always tried to take the used flight feathers before he could ruin them, because they were treasures to me. But this time, Wesley was one step ahead. He held the old feather out in his beak. Surprised, I took it and thanked him, making much of his gift, and placed the feather in a vase where he could see it. But Wes wasn't finished. The equal and opposite feather had to come out, too. Wes swung around to his other wing and presented me with the matching flight feather, his deep obsidian eyes locked on mine.

Grandpa still played and taught drums and continued with his life. After an appropriate time had elapsed, the predictable "casserole brigade" started appearing at his doorstep. Unfailingly polite, he thanked each lady for her kind attention, but told her there was only one woman for him and that was Grandma. He had lived his lifelong love. This was the Way of the Owl.

THE YEARS I spent in labs and lectures and reading textbooks deepened my view that the universe is a place of wonder and meaning. Science has made many thrilling discoveries, but along the way it has also opened up myriad, endlessly branching questions. It's like we are scrabbling in hard dirt with our hands, trying to reach China, and have barely broken the the surface. Many scientists consider the idea that there may be something more that science will never be able to explain. At Caltech, a sizable group of physicists felt this way, some with Nobel Prizes. The more they gazed into the vast stretches of the universe, or the vast empty spaces within atoms, the more wonder they felt. They formed a group that met once a week to discuss the spiritual side of their experiences.

I gazed into the universe of Wesley's eyes almost every day for nineteen years. We communicated—spirit to spirit. I can't really explain it, but things occurred to me when I "listened" to him, thoughts that were not my own. Perhaps he was as thoughtful as I, but in a way that I could never touch or understand; perhaps he understood and saw things that I can't. When I would look into his relaxed, at-peace-with-himself eyes, I felt like I was looking into something inscrutable, unobtainable, deeper than we can possibly imagine, an old soul that reflected something bigger, ineffable, eternal.

Even though I had been trained to exclude thoughts about spirits and unquantifiable, immeasurable feelings that could taint scientific conclusions, Wesley's presence in my life influenced my thinking. Now I see that to exclude a certain kind of idea is itself creating a bias. What if the truth screams as loudly as a male barn owl crying for a mate, and we miss it because we have not allowed ourselves to listen to the channel it's on—or we've tuned it out? Wesley helped me feel God—to "get" the idea of God and the soul in a way that I had not before and couldn't get from a theological sermon. I've decided not to discount those feelings and the wonder and gratitude that comes with them.

13

The Sex Tapes

ONE OF MY favorite things about living in my new town was the local coffee shop, La Costa Coffee Roasters, established long before a Starbucks anchored every neighborhood shopping mall. Their breakfast blend is beyond compare. Plus, they had a gift shop filled with owl merchandise. It was almost like being at Grandma's house. I bought a stuffed owl one morning and mentioned to the owner, John, that I was an owl fanatic.

"In fact," I said, yelling over the bean roaster, "I have been taping wild owl sounds for several years now." I took a chance that he might be an owl buff, too. "So if you ever hear about any nesting owls around here, let me know and I'll tape them."

"The owls? Okay, that's fine," he yelled back.

"No, not toy owls, real owls," I shouted.

"Yeah, they're here," he yelled again.

"No, I mean living owls, like a nesting pair."

"Yup, they're still here," he said again, looking annoyed.

Having trouble getting through to him, I clarified: "No, John, what I mean is I'm looking for a pair of barn owls that I can watch at night and tape-record."

He turned off the machine.

"And I'm telling you that, yes, we have real barn owls. Why do you think my shop has all this owl stuff?

"You . . . you have an owl nest?" I said, incredulous.

"Yup. They've nested here for four years now. And they're so noisy that the first year I thought I'd lose business. But the weirdest thing is no one even notices them. People eat outside while the owls fly right over their heads and scream. No one even looks up. On Saturday nights we have live music out there and the owls almost drown it out, but nobody says a word. People are strange."

I couldn't believe my luck. "Could you show me where exactly?" I asked.

"Sure." He wiped his hands on a towel and we went out the side door.

He pointed to the top of a decorative stucco tower.

"The nest is up there under the tower roof," he said.

I had struck gold. I could park my Toyota Celica right below the tower next to the coffeehouse and have a clear view.

"It's a good thing the cleanup crew gets here really early," said John. "In the morning there are mouse parts all over the place and these brown fur balls. What are those things anyway?"

"They're owl pellets." I explained how owls digest their mice and cough up the bones and fur.

"That sounds miserable . . . like a cat choking up hairballs," he replied. "I had no idea, I just figured owls pooped a lot."

We went back inside and he turned the coffee roaster back on.

"Have a good day! Thanks so much!" I yelled.

"Good luck." He waved.

I couldn't wait to visit these owls, so that evening I took a nap and drove down to the coffeehouse at around 2:00 a.m. By then the movie theater had let out and the parking lot was empty. I rolled down my window and, sure enough, heard the unearthly din of five screaming barn owl babies clustered on top of the tower roof with their mother, begging for food. The oldest and most developed owlet began hopping between the tower and the roof of the main building, with an assist from fledgling wings.

The father owl was hunting frantically, shuttling back and forth from the nearby open field, each time with a mouse in his beak. He was having no trouble finding mice, but the babies were so ravenous he could barely hunt fast enough to please them. *Poor guy*, I thought. *I bet he's just exhausted by morning*. I opened my sunroof and set the tape recorder on top of my car. The owls were used to cars, so they ignored my presence.

The father was noticeably smaller than the mother, with mostly white feathers underneath. Female barns owls are usually darker than the males, but that's just a rule of thumb, so it can still be hard to tell them apart. The main difference is size. The female is about a third larger than the male and much more aggressive. In most of the nests I've observed, the mother is a screaming banshee—hell on wings—and the father is a "dear old dad" type—pretty laid back and mellow. Their personalities make sense. The female has to defend the nest and the male has to stay focused on hunting and not get easily frustrated. If he weren't mellow, he'd have abandoned the screaming bunch a long time ago; but he's patient to a fault.

The father continued to hunt while the babies and mother screeched and jockeyed for position on the tower, anticipating his next delivery. I was watching this owl family drama for a couple of hours, recording a fascinating variety of vocalizations, when a delivery truck pulled into the parking lot. The sun had not yet risen, though the dark sky was softening on the horizon. The driver began unloading supplies for the coffeehouse and carried a box across the outdoor eating area to a locked storage room.

The mother owl was already worked up, and now her territory was being invaded. She chirped to her babies and they hunched up next to her, silently. Then she stood on the edge of the roof, gathered herself, and dived straight down with a shrill scream. It was like no sound I'd ever heard from an owl, more like the scream of an enraged eagle, and even from a distance it about burst my eardrums. At the last moment she pulled up, both talons spread out with claws bared, and raked the air over the man's bald head, missing by an inch. Then she flew back up to the top of the tower.

The man didn't even notice! He didn't flinch, didn't look up, didn't change his bored, trudging stride. Nothing.

The mother owl was as perplexed as I was. Perhaps she needed to do it again. And this would be a good lesson for the owl babies, who had stopped their constant roughhousing and screeching to watch her intently. So she tried again, producing an even more vicious screech. She slicked her body tightly for speed and dived straight down, pulling up at the last second, talons raking over his head, missing by just a wisp of air. Again he paid no attention at all!

By this time, I was pressing my face into a pillow, trying to muffle my laughter. I didn't want to mess up the recording.

It never crossed my mind that the deliveryman was in danger. The mother owl just wanted to scare him off, though it certainly wasn't working. He continued unloading supplies. Unless that deliveryman was deaf, he had to have heard those earsplitting screeches. And how could he not have noticed the air currents from a rather large object that was practically flying into his head? Somehow, the guy had completely tuned out his immediate environment. For him, this was a tedious morning like any other that he needed to forge through.

Shortly after the deliveryman left, I packed up my equipment and pulled out of the parking lot, exhausted but happy with the success of my recording session. I hoped Dr. Penfield would find some interesting and unusual owl sounds. I was sure he'd heard every vocalization wild owls could make, but for me it was a thrill hearing them firsthand. I still had much to learn from barn owls.

When I got home Wesley greeted me with his chatter of unfettered joy. He certainly knew how to live in the moment. Never jaded, he was always so happy and boisterously expressive whenever I returned from my short trips away. He would update me with chirps, twitters, long patterns of exuberant cries, and even tiny hisses when he recalled something that had happened during the day that he didn't like—such as someone besides me feeding him. In that instance, he'd stare at the place where that person had stood and hiss while recalling the hand that fed him. Then he'd go on with his daily update. If he had been sleeping, he'd turn, hop up to the dowel, then do a long arabesque with one wing and leg stretching back.

I had the tape recorder with me so I popped in a fresh cassette and left it running while I unleashed Wesley. He played all over the room, chattering and commenting on this and that. He always "talked" about almost everything.

"Okay, Mr. Constant Comment," I told him, "you sure have a lot of good things to say to Dr. Penfield's machine tonight."

Wesley flew up to the top of the curtain rod then down to my bed and upon landing, bent over the quilt quizzically, looking under each fold to see if there were any surprises. Perhaps he could even wedge himself in there. He poked his head into the fold. No, his body wasn't going to fit. He cried, "Reek reek reek reek!" and leapt into the air, pouncing on his personal pillow. He then looked over at me with great interest as I sat on my side of the bed.

"Bzzzzztttthhhhh," he screeched (some people say it sounds like a handsaw when owls make this quiet screech), asking for a magazine.

"Do you want one of these magazines, Wes?" I asked.

"Twitter twitter" was his answer, i.e., "yes."

"Okay, here you go."

He started ripping it up. After a while he got tired of this and wanted to cuddle, so he came over to me with soft twitters and chirps, some barely audible. I still had the recorder running. He and I talked quietly back and forth, his soft sounds recorded for the first time. Then he climbed into my arms and fell asleep, lying as always on his stomach with his feet dangling, head in my left hand. As I stroked his neck feathers and ears, he made such tiny sounds that I couldn't actually hear them. I could only feel his diaphragm moving. I answered with an almost equal softness to let him know I'd "heard" him.

Over the following week I spent the wee hours of every morning at La Costa coffeehouse tower recording the wild barn owls. The owl babies soon began taking short flights to nearby trees, sometimes settling there for the day. Now customers were starting to notice the owls and to mention them to John. They were so beautiful that they began to attract a small audience of bird watchers during business hours. But between 2:00 and 5:00 a.m. I had them all to myself.

I decided to venture out of my car, since the owlets were now flying all over the place and getting hard to see from only one location. I was careful to move slowly, and for whatever reason, the owl mother didn't harass me, maybe feeling that by now her babies could fend for themselves. But I could see that the father was running out of strength, looking peaked and downright unhealthy. He was still hunting constantly with the pressure of owl babies screaming at him, only now they were actually attacking him when he flew into the tower with a mouse. They had grown as big as he was and he wasn't taking it very well. He couldn't keep up with their demands.

I thought of the big bag of frozen mice just up the hill at my apartment. I just couldn't stand watching him exhausted and frazzled, hunting like mad only to get pushed around every time he brought food home. He was so faithful—trying so hard.

I drove back home, grabbed a big bag of mice, and defrosted enough to feed the entire owl family for twenty-four hours. *This should give Mr. Owl a break,* I thought. I returned to my parking spot and got out of the car with the bag. If I could just throw the mice, one by one, up to the top of the roof, then surely the babies would eat their fill and the father could have a night off. I took one mouse, wound up my arm, and threw it with all my might. It fell painfully short of the top of the tower. I had never been good at this kind of thing. I tried again and again, but there was no way to get any pitching power out of my arm.

So, here I was at 2:00 in the morning at a shopping mall throwing dead mice into the wind like a crazy person. I had become so intent on my task that I was completely startled to discover myself surrounded by a group of teenagers clad in black leather with piercings, tattoos, shaved heads, and muscle.

I felt very alone, very small, very blond, and scared out of my wits. They were just staring at me.

Finally, the biggest among them said, "What are you *doing?*"

"Uh . . ." My heart was in my throat. I pointed up. "See those owls?"

"No, I don't see any owls."

"Well . . . there's a nest of barn owls up there. And the father, uh, he's exhausted because he's been hunting and hunting for a couple of months, every night, and the babies are overwhelming him. He can't hunt fast enough to keep up with them. See, if you look right up there, that's him."

I pointed again.

The tough kid looked up, "Oh! Okay, yeah, I see him."

"And you hear that loud screaming? That's the babies."

"*That* sound is baby owls? No f____ing way!"

He was interested, to my surprise. Perhaps this was not my last night on earth after all.

"Okay, there's a baby in that tree over there, see, and the rest are on top of the roof." All of the kids craned their necks to see.

"And I've been trying to throw these mice up there to give the dad a break, but I just don't have a throwing arm and, well, there are mice all over the patio now 'cause I just can't get them up that far."

For the first time they looked around and saw dead mice littering the concrete, then the big plastic bag full of dead mice, and finally, the four I had in my hand.

"Oh, *dude*. This chick is throwing *mice* up to these owls! No *way*, man!" I could hear them all commenting to each other "Rad!"

The ringleader stepped up and said, "I can throw. Give me those mice."

Amazed, I found myself placing four cold limp mouse carcasses into his huge palm. He hurled each one with pinpoint accuracy right up to the top of the roof, and the owl babies pounced on the mice and powered them down. Then he gestured for the bag and I handed it to him. The rest of the kids started throwing the mice that were lying around on the concrete.

"Gross! Wow! Get that one! Get that one, Mike! Yeah! Got it! All right!" They gave each other high fives every time a mouse was on target. Then they'd watch the baby owl devour it. "Look at that guy eat that thing! He's eating it whole. That's rad, man. Cool!"

"Dude, I've never seen real owls before. Do they only eat mice?" asked the big kid who had first approached me, then he stopped short, wondering the obvious. "Where did you get mice at two in the morning? What are you doing with a big bag of them?"

I explained that I had been studying owls all my adult life and had majored in biology and worked with scientists.

"Man, that's what I want to do!"

They threw all the mice until the owlets were sated and sleepy. The mother finally calmed down, and even the father was full of food and relaxed. The little owl family didn't seem to mind our presence at all.

Sitting along a concrete wall, the kids and I had a fascinating discussion about owls and their ways. Then we gave each other high fives, and they melted into the night. I was left standing there beneath the tower as the early-morning ocean fog rolled in, watching the owls resting with full stomachs and enough mouse snacks to last well into the following evening.

DR. PENFIELD E-MAILED that he needed his recording equipment back. "Bring any tapes you have made so we can listen to them," he said. We scheduled a day for me to visit his office at Caltech. Living two hours away, it'd been a long time since I'd dropped by. I missed the people and atmosphere there and was looking forward to an opportunity to reconnect.

Because Dr. Penfield was so respected in his field, I was intimidated to talk with him, but he was always gracious.

"Have a seat," he said, on the day of my visit, pushing aside piles of papers and books.

I set up the recorder and started playing the screams of the wild mother owl protecting her babies. He was fascinated by the range of sounds and asked me to narrate the accompanying owl behavior. I then played Wesley's intimate vocalizations as we snuggled and played with magazines, as well as Wesley's chattering "updates" when I came home from work. Dr. Penfield peppered me with questions and scribbled notations.

"What else do you have?" he asked.

"Well, uh . . . well, here's one with Wesley's nesting call. Although, I'm sure you've heard that in the aviaries at night, right?"

I handed him the tape and he inserted it into the player and listened. We went through the foot stamping and "re-EEK re-EEK re-EEEK" of his nesting calls.

"Okay, he's found a nesting spot, and has possibly added his own improvements to it, such as a ripped-up magazine," I explained.

"So you know for sure that this is a male?" he asked.

"Yes," I answered. But he patiently suggested that I might not be able to be as certain as I thought I was.

"Oh, I know he's a male, Dr. Penfield," I said.

He lifted an eyebrow, then continued the tape. "Okay, go on."

"Well, he's decided that the nest is ready for eggs, so he's calling his mate . . . uh . . . ahem . . . he's calling me to the nesting site . . . you see . . . he thinks I'm his mate . . . [*cough cough*]."

If Dr. Penfield hadn't been such a distinguished gentleman, it might not have been so difficult, but he had this aura of . . . well, of being so darned distinguished. Everyone felt it. Sitting in his office was like sitting in church.

"He is bonded to you as a mate?" Dr. Penfield raised both eyebrows this time and paused the tape. "One moment, Stacey."

He slipped out of his office and to my horror called in several postdocs. He rewound the tape.

"Good morning, everyone. This owl, Wesley, has bonded to Stacey here as his mate." He gestured to me and I managed a weak smile. He hit Play. Wes started his nesting call again, but then the sound changed and Dr. Penfield said, "Now, what exactly is he doing here?"

"Well, uh, this is the part where I come up to the nest." The office felt very hot and stuffy.

Wesley was making little awking sounds of familiarity and affection. Then there was a more urgent "ur ur ur . . . ururu-rAWRk urk urk AAAWWRRRK . . . ur ur ur ur . . . ur ur ur AAAAWWRK."

Dr. Penfield's eyes widened and he paused the tape again. "What is this sound? *I have never heard this before!*" he said.

"Well . . . he's positioned himself over my arm and . . ." More postdocs began to arrive. I heard whispers in back of me. "Have I missed anything?" "Yeah, the owl thinks she is his mate . . ."

"How?" Dr. Penfield continued. "How is he positioned over your arm, exactly?"

"Okay . . . well . . . uh . . . Okay . . . let's see . . . ummm . . . he's holding on to my left arm with both of his feet, gripping with his talons. I have my right arm perpendicular to my left arm, between his feet—he's grabbing hold of my right arm, uh, with his knees . . ." (There was a titter from a postdoc.) " . . . The first time, I tried to fight him off, but it always becomes a physical battle unless I just let him go through with it . . . and . . . er . . . well . . ."

Dr. Penfield was growing more and more enthusiastic. I was becoming less and less enthusiastic.

"Oooh! So your arms are like this and he is like this," he positioned one arm where my arm would be and his hand where Wes would be. "Am I right?"

"Yeah, that's it." The postdocs leaned in. I slid down into my chair wanting to disappear.

"And then what is this repeated sound?" He asked.

"Uh . . . well . . . uh . . . that's . . . well each 'ur ur ur ur' indicates him dipping down onto my arm and clutching with his knees and, well . . . um, rubbing on my arm sort of, and, well . . . you know." (I heard increased chortles of delight behind me.)

"*Aaahh!* I seeee! So those are the 'ur ur ur.'" He imitated the sound and indicated the movement of the owl on my arm by using his hand on his arm.

"Yes," I muttered in agony.

"Well, then," he continued, "What is the final AAAAWRK AAAAAWRK AAAAWRK sound then?"

"Okay, well . . . that's the actual, the . . . the . . . well . . . the actual . . . I guess you'd say it's the . . ." ("Hee hee hee" wheezed a voice behind me. I wanted to punch the guy.) Now the boss was just beside himself.

"Do you mean to say he actually has an *orgasm on your arm*? How do you know? Does he ejaculate? Is there actual sperm?!?"

"Oh, yes," I managed to say, "there's sperm. He ejaculates, all right."

"How much sperm? Is there a lot? How do you know it is sperm? Did you look at it under the microscope?" (The voices behind me were now choking with glee.)

My answers came out in rapid fire.

"About one-eighth of a teaspoon. Yes, I looked under my microscope, and it is definitely sperm."

Dr. Penfield slapped his desk. "These are new sounds! And no wonder you knew he was a male!" He started delegating responsibilities to the postdocs. One ran to get the equipment. Another loaded the tape into the computer, and they all sat around while the contents of the cassette slowly appeared on the screen as sonograms. As each one came up, I explained every *urk* and *ark* and *ur* and *de-Deep* and mutter and grunt and chirrup, while another PhD sat taking notes madly.

One woman asked if she could use my data for the PhD she was working on. Since I didn't intend to pursue a PhD, I gave her permission. There was loose talk of my perhaps getting all of this on video and bringing that in, too, but it wasn't followed up, thank goodness.

At the end of the day Dr. Penfield told me, "Stacey, perhaps the most interesting aspect of your tapes is that you have fifteen or twenty different complete sequences of Wesley's mating vocalizations, and none of them is exactly the same. If his mating act were purely instinctive, then the sound sequences would be identical every time, like a birdsong. But there is so much variation in his expressions that one has to conclude that each sequence is an individualized experience from the owl's point of view."

I beamed. This was additional confirmation that owls are anything but simple. Wesley and other owls are emotional and show their feelings. They are intelligent and communicate their thoughts in creative ways that we don't always recognize. In fact, many of the higher animals are not really that different from us, they are just "other."

14

Fifteen Years of Trust

BACK BEFORE JOINING Caltech's owl lab, I had thought owls only had one or two "looks," but Wesley could move the skin under his feathers to create countless expressions, directing little groups of feathers this way or that. Each variation meant something vastly different. Sometimes he looked like a quizzical butler wanting to help in any way he could. At other times he looked like a Buddhist monk who had obtained the highest level of enlightenment. His black eyes could express mischief, ferocity, love, gentleness, innocence, intelligence, awareness, and absolute trust. When I paid attention to his facial and body language I could often discern exactly what Wesley was thinking, feeling, or about to do. It got to the point where I could "read" his feathers.

A slight lifting of his third primary flight feathers meant he was thinking about flying. If he flattened the feathers around his face and dipped his neck slightly forward, it meant he might actually fly. If he spread his feet slightly apart, gripped his talons a little tighter, and focused his face sharply toward a landing spot, it meant he had, in fact, plotted a trajectory along

While grooming, Wesley can pull patches of feathers into position for better access by moving the skin underneath. *Stacey O'Brien.*

which to fly. All of these were stages that gave plenty of warning and preceded the more obvious mannerisms of bending his legs and lifting his wings.

"How did you know that owl was going to do that?" the other workers were always asking me at the wildlife rehabilitation center where I volunteered.

"You get at least three subtle warnings before they actually lift their wings to fly," I would reply.

Not only did Wesley seem to have infinite physical expressions, he kept up a running verbal commentary of twitters and chirps (positive), and hisses and clicks (of disapproval) on everything around him. From the day we moved into our own apartment and had our home to ourselves, he followed me everywhere, like a tiny man walking three feet behind, just tagging along to see what I would do. I loved my elegant little shadow, chattering constantly to himself and me. Like a play-by-play sports commentator, he'd describe everything he saw and like a color commentator he also conveyed exciting nuances with plenty of animation and exclamation. Amazingly, after fifteen years, I could understand most of it. He was the narrator, and I was the star of his show.

She's going into the kitchen . . . a fascinating place with all the clangy bangy shiny stuff, but it's not for owls, no, no, no, not for owls and I'm not allowed in.

One of the joys of having our own place was that I could bring him into any room, so that he could finally see what I did in the rest of the apartment. I'd set his perch so he could watch me in the kitchen from a safe distance.

"I'm on my tether because she's going to cook . . . and OOOOH she's gonna make spaghetti! What an afternoon! She's off to the sink. Water!" Eedle-deedle-DEE-DEE-deedle-deedle-dee!" *Oh dear oh dear . . . "szzzhzhh," I don't approve of the stove . . . Oh, "hiss" and "click." She tells me it's dangerous so why does she go there?*

Oh, no. Steam. I don't like that hissing sound it makes so I'm going to hiss. "HISSS."

I sat at the table and began to eat my meal. Wesley started pacing and picked up a mouse off his perch tray.

What are you eating? That's disgusting. I must intervene! I have a mouse for you! Here! A mouse, I say, a mouse! Look, look, look! Eat it!

He lunged at me urgently with the mouse in his beak.

Ugh, how can you eat that gross spaghetti? "Bbbrrrrzzzzztttt." *Oh, gag, "hiss," yuck.* "Click click."

As Wesley settled into a comfortable spot, his eyes would start to get heavy, then the sky-blue nictitating membranes would close partway, or he'd leave one eye slightly open and the other would droop comically. He'd fluff his feathers and transfer his weight onto one foot or the other, getting cozier and cozier. Then he'd spread out his forehead and nose feathers until his face assumed a squashed appearance. At some point he'd decide that he really was going to go to sleep.

I'm tired. I think I'll have a short nap.

He fluffed and began to pull up his leg.

. . . Oh, wait . . . What was that animal sound and where is the animal? A hamster is out! Against the rules! Against the rules! "Twittertwit hiss. Click click click."

Oh, good, she captured it and returned it to its rightful place. Big sigh.

Oh, my goodness. What a wonderful life this is! What a wonderful incredible life!! Everything is sooo interesting.

But now I really want to take a quick nap. Okay just a little nap. Take a little nap take a little nap. "Teep deep deep teeple teeple teeple . . ."

Oh I have an itch. "Hiss." *I hate that just when I'm settling down. Okay, now I'm about to go to sleep.* "Teep deep deep teeple teeple teeple."

He fell asleep and had a dream and screeched.

. . . AAAAAAAA! "*Who was that? Did you screech? It wasn't me! You woke me up! Why did you screech?*"

Now he was talking to me directly. Wesley was always outraged when he woke himself up with a screech in his sleep, and he blamed me. He would whip around to face me with an intense librarian's stare as if I had broken a cardinal rule.

A relaxed pose on his adult perch with one foot halfway up. *Stacey O'Brien.*

Someone screeched. I about jumped out of my talons! I hate that! Did you do that? Did someone outside screech?

Then he would shake his head and look around the room.

He turned in circles, then decided there was nothing to see.

Oh well, I guess I'll go back to sleep. "Teep deep deep teeple teeple teeple . . ."

I knew he was finally serious about taking a nap when he'd pull up one foot until it disappeared into his tummy feathers. That leg would now bend sideways, resting on the crook of the other leg, much the way we cross our legs, though in the opposite direction. Then he might do a big final fluff, heave a deep sigh, and sink down onto his little leg platform. Sometimes if he were fighting sleep, his eyes would start to close, but then he'd keep looking over at something he wanted to investigate. His free foot would go slowly up and slowly down, up and down . . . the restless leg of indecision!

Wesley looked so cute at these times that I would often have to pick him up and let him snuggle in my arms. Of course, then the atmosphere would become so soporific that soon I would lean back on some pillows and we would both doze off together.

A WILD OWL would rarely live to fifteen years. I didn't want to think about Wesley getting old although I knew it was inevitable. One night I heard a loud thump in the bathroom and found him rolling around on the floor silently, panicked, having fallen from the curtain rod. At first I couldn't figure out what was wrong. I knelt down beside him and saw that his talon, which had lengthened and curved as he aged, had impaled the flesh of his wing—the weak one that sagged a little. He lay on his back trying to yank the claw out, but every time he pulled at it, a new wave of pain shot through his wing.

I tried to unhook him. As soon as I took hold of his wing, Wesley grabbed my hand and sank his meat-shredding beak through my flesh, the ends of his beak meeting inside my hand. As we both screamed, I worked frantically to extract his talon from his wing. When I finally succeeded, he let go of my hand. I lifted Wesley into my arms, where he lay quietly while I gently explored his wing for broken bones. Thankfully, there were none.

When a wild animal is alone and in trouble, his instinct is to remain silent so as not to alert predators that he is injured and vulnerable. Once help comes, however, he feels less vulnerable and may finally scream or bawl, giving voice to his pain and fear. This is what Wesley had done.

I held Wesley all night, comforting him, because I worried he'd go into shock and die like the owl who got lost in the Caltech ventilation system. He lay across my arm and against my stomach, his injured wing hanging down, and I stroked his back and neck, speaking softly to him as he slept on and off. By morning he seemed to feel much better and was back to his old self. I was an emotional wreck, however. My hand was swollen, but since owls have such acidic saliva, his beak was extremely clean, so infection was not a concern.

At first I thought this had been a freak accident, but he did it twice more in as many days. I had promised myself I would never clip his talons because that would dull the needle-sharp tips and I wanted him to have his full defenses if he ever needed them. But now instead of serving him, his talons had become a serious liability. I would have to trim them.

As soon as I approached Wesley with a talon clipper he panicked, leaping wildly into the air and flying in aimless frantic circles, searching for a high place to land. I wondered if I should chase, catch, and forcibly hold him. I had a lot of experience restraining wild birds, particularly owls and ocean birds, in order to help them in times of injury and trauma. But restraint was tough on them. It often took two people, one to wrap the animal in a towel and cover his eyes and another to do the procedure. I didn't want to treat Wesley as a wild animal and risk losing his trust. I was afraid I'd never get it back. An owl never forgets anything. Ever.

Before I could even solve this issue, another problem arose. Wesley's beak was growing longer, sharper, and more curved at the end. He accidentally embedded his beak into the skull of a mouse and because it was now so long and curved due to old age, he was unable to pull it out. He started to choke because the mouse was stuck inside his mouth, partially blocking his glottis, and he tried to grab the mouse with his feet. His breathing became raspy. I could see that Wesley would not be able to sort this out himself, so I took hold of the mousehead in one hand, Wesley's beak in the other, and pulled them apart. It was quite an effort. I doubt this happens in the wild, but it was most likely a complication of old age that would be extremely rare. I'd have to figure out a way to trim his beak as well as his talons.

How would I ever convince him to allow me to approach his beak with a metal file, which would flash in his eyes and vibrate his head? I also feared clipping his talons, as he hated to have his feet restrained for any reason. If we were cuddling, I could gently stroke his feet and play with his toes, which he liked, but that was different from holding them tightly in place while accosting them with a shiny, clacking object.

My first attempt was to slip the clippers under a talon while he was asleep in my arms, but he was far too quick for me. He must have sensed something afoot because he flew straight to the highest point in the room. I couldn't coax him down. Remaining suspicious for the rest of the evening, he wouldn't relax.

Next I tried sneaking up on him and grabbing a toe, quickly positioning the clipper and clamping down, while holding fast to his foot. He strained and flew against me, fighting like crazy to free his foot. This made it impossible to trim the talon without cutting into the live part of the claw, which would have hurt and made him bleed profusely.

Finally, more out of desperation than cleverness on my part, I began to work with Wesley using language and imagery. Some scientists believe that animals may use some sort of mental telepathy to beam picture thoughts to communicate with each other, and experiments indicated that it does work between humans and certain animals. It seemed pretty far out, but I decided to try. I sat still and began to send thoughts and pictures to Wesley about trimming and filing. I also spoke the thoughts out loud because he was used to my talking to him in a deliberate manner before introducing something new.

I decided to focus on his beak first because, until that problem was fixed, I had to cut the heads off of all his mice before feeding him, or he would choke. So the next time a mouse got stuck on his beak and he was distressed, I said, "Wes, your beak is too sharp. Your beak is stuck. Let Mommy fix your beak . . ." over and over again. I took the mouse off his beak and continued to talk about it. He knew "Mommy," "fix," and "beak," so it wasn't too far-fetched to conclude that he could string those words together in relation to this problem. Then I would visualize an image of me peacefully filing his beak.

I showed him the file. I filed his dowel perch, I filed my nails, I held his beak lightly with my fingers and said, "Mommy can fix your beak with this." He jumped away and flew in circles. I didn't pursue him.

I continued this process for a few weeks. Any communication with an animal takes a lot of patience, but building trust takes time, and it's the only approach that works long term.

One day I said, "Mommy is going to fix your beak in two days, okay? Two days, Wesley. So think about it and get ready. I will not hurt you, but I will fix your beak in two days." Then I'd beam visualizations to him of a very peaceful procedure. At different times during the two-day period I would approach him with the file and he'd panic. I'd drop back. But then I'd tell him how long until I was going to fix the beak. "Mommy will fix your beak tomorrow, okay, Wesley?" Meanwhile, I used the file around the room on other things.

The day came. "Okay, Wesley, Mommy will fix your beak in two hours!" I told him, always mentally sending him the image of myself peacefully filing his beak.

Then the moment arrived. I slowly approached his perch. "It's time to fix your beak now, okay?"

Wesley closed his eyes, hunched down, braced his legs, and stood perfectly still. I was amazed. I began to talk gently to him as I grasped the top of his beak, then started to file. He pulled back just a tiny bit and squeezed his eyes shut. I filed and filed. It seemed to take forever to trim that long ice-pick-like hook back to a manageable size. Wesley didn't move a muscle or make a sound. He just kept his eyes shut and acted like he was intently focused on not feeling anything. The filing must have vibrated his head and made a terrible sound inside because the beak is really part of the skull.

When I was finished, I wiped his beak and said, "Okay, Wesley! Good job! All done! What a smart bird! So brave!"

He seemed relieved and very pleased with himself. I unhooked his tether and he leaped right into my arms and relaxed across my left arm, his feet hanging down, and went to sleep for a while. He was exhausted by the effort of letting me do this. But the point was that *he* had chosen to let me do it.

I used the same approach with his talons. On the appointed day he presented his feet for trimming. Again he looked away and, with utmost concentration, tried to ignore what I was doing. He closed his eyes but relaxed his foot and let me carefully cut and file his claws so that there were no raw edges to catch on anything.

What a relief.

I couldn't wait to tell Wendy about this on the phone. She revealed that she'd also been using this method with her horses with fantastic results. One horse in particular, named Chica, was terrified of the horse trailer. Because Wendy was moving soon, she would have to haul the mare six hours to the new place, so she spent some time sitting quietly with Chica talking about and visualizing images of the new property. When the appointed day came, Chica walked right into the trailer without Wendy even leading her. Wendy was astonished. "It really works!" she told me.

People working with all kinds of animals are altering their methods from those that used force and negative consequences, like spurring, hitting, shocking, or yelling, to gentler approaches of positive reinforcement. Horse whisperers are explaining their gentling techniques. Zoos like Steve and Terri Irwin's Australia Zoo are encouraging relationships between the animals and their keepers. Scientists are teaching language to parrots and sign language to chimpanzees. Many scientists now are not interested in mere "behavioral modifications," but instead see their interactions with animals as flowing from a true relationship with the animal—a partnership with a fellow sentient being. These interactions are infinitely more complex than mere behavioral modification. Do humans consciously use a behavioral modification technique to teach language to their children? No. The learning flows from a loving, mutual relationship. I do not see most animal learning as being about behavioral modification. Wendy says that it's her relationship with a horse that results in mutual cooperation, that is, actual friendship. Friends do not "modify" each other's behavior but they do teach and learn from each other. This goes far beyond "training techniques."

Some researchers are also accumulating empirical evidence that animals use a form of telepathy to communicate with and understand us. Recently, Jane Goodall, who seems always to be one step ahead of everyone else in animal behavior, hosted a Discovery/Animal Planet documentary showing some of the latest experiments that demonstrate that animals use telepathic communication. Several experiments showed that some dogs can tell when their owners are about to come home, even without the cues that people had thought the animals were associating with their arrival, such as the sound of the car, the time of day, or footsteps.

The most impressive experiment, to me, was one involving an African gray parrot who had a large vocabulary and chattered to himself constantly. The owner was set up in a completely separate building, far from the parrot, and given a series of cards that neither she nor the parrot had ever seen. There were two cameras—one on the parrot and one on the owner, with a timer running. Then the owner picked up a card and looked at the picture on it. It was a blue flower. The parrot, at that same time, began to talk to himself about blue flowers, pretty flowers. Then the owner picked up a picture of a boy looking out a car window and the parrot's chatter changed to "Do you want to go for a ride in the car? Watch out. The window is down. Look out the window." I am paraphrasing, but the conclusion of the experiments was that animals and humans were using telepathy.

Many pet owners already believe this, and certainly Wesley and I benefited greatly when I opened myself to using my own intuition to understand him. When humans and animals understand, love, and trust each other, the animals flourish and we humans are enlightened and enriched by the relationship. Wendy has a magnificent black Friesian stallion. Her way with horses is to establish a deep mutual bond of love and respect out of which flows polite behavior from both her and the horse. The horse hugs her by putting his head over her shoulder and pulling her into his chest, her arms wrapped around his neck.

In today's technological world we have lost a great deal of ancestral knowledge of animals and nature. Many people can be said to have a nature deficit disorder—an estrangement from the natural world and their own basic nature. This intuitive mode of communication may have been very familiar to our ancestors. I like to think that we're evolving as we learn—or relearn—how much more complex and intelligent animals are than we've previously admitted and how deeply connected we all are.

I could have forced my will on Wesley, and it would have destroyed the trust between us. Because I took the time to communicate with him, he realized that I wouldn't do anything to him without asking him first. I had allowed him to be part of the process and to maintain his dignity. Our relationship changed. Going through this together awakened a deeper bond of trust.

WESLEY WAS CLEARLY aging, so I knew it was time to find a good vet who could work with him if he had a crisis. It had to be someone I knew and who understood captive wild animals. This vet would also have to be willing to keep knowledge of Wesley limited to a few trusted people.

I already had an amazing vet for my hamsters, Dr. Douglas L. Coward in Mission Viejo, who specialized in exotics and wildlife, working with animals all over the world: elephants in Nepal, tigers in Thailand, apes in Africa, marsupials in Australia. We had developed a mutual respect; we both related to animals and cared for them. Dr. Coward has a spiritual side and makes every attempt to quiet himself and "hear" the animal. A true healer who looks beyond physical symptoms, he brings a holistic approach to his patients, treating mind, body, and soul.

Once Dr. Coward diagnosed a hamster with Cushing's

disease—a rare disorder caused by a brain tumor. I asked him how on earth he was able to make such a diagnosis, which requires hormone tests, MRIs, and CAT scans. Dr. Coward was so humble he wouldn't have mentioned it if I hadn't asked.

"Well, to tell you the truth, I did involve a hospital," he said.

"You mean a hospital for humans?" I asked.

"Yeah, I thought your small buddy needed more expertise than I had here so I called a friend of mine who's an endocrinologist and he came over on Sunday and we spent the day diagnosing your hamster."

"Are you kidding? Don't those guys cost four hundred dollars an hour?"

"Usually, but this was his day off and he did it as a favor."

That was typical of Dr. Coward. He'd go to any length to help his animal patients. I have never seen such dedication in a veterinarian.

I told Dr. Coward all about Wesley and asked if he would be willing to be involved, and whether there would be a problem with anyone in his building if I brought in an owl.

He said, "No problem. I treat owls all the time and love them." I told him I wanted to keep Wes a secret and he said, "Okay, call in and use a code. Just say, 'I need to bring the bird to Dr. Coward,' and ask them to treat it as an emergency."

Knowing Wes now had a great vet, I had some peace of mind.

After Wes and I survived the beak and claw crisis, life returned to normal. Yet having triumphed over a great difficulty, we grew even closer. Throughout our years together, we had a cuddle ritual almost every night before bed. I would scoop him up with one hand cupped under his tummy, then cradle him in my left arm with his head resting in my hand. He would immediately pull up his feet like landing gear, letting me swing him gently into place against my stomach.

One evening, however, as I was lying down and rubbing him under his wings, Wesley pushed with his feet so that he was lying on my chest with his head up under my chin, his beak sleepily nibbling my throat. Then he rustled a bit and slowly began to open both delicate golden wings, stretching them as far as they would go and laying them across my shoulders. He slept that way for a long time and I stayed awake in awe.

It was an owl hug. I hoped he would do it again. He did, and this vulnerable position became his new way of cuddling. I never got over the wonder of it and I often felt tears stinging my eyes. This complicated wild soul had stretched his golden wings over me in complete trust. I wouldn't trade those moments for anything in the world. Not for anything in the world.

15

Twilight: He Whom I Tamed Saves My Life

MY WORLD WAS about to change dramatically. It was 1998, I was in my late thirties, at the top of my career, doing well financially, enjoying my friends and family. Everything was great. But one day I woke up lying across my front door threshold. As I gathered myself, I realized I'd blacked out and been unconscious all night. There was no explanation—I had no injuries, I was not a drinker, and I had never used drugs. Thankfully, nothing was missing from my apartment, so I knew I hadn't been robbed. It was altogether baffling, but I continued with my life.

Then I passed out on a Sunday night and didn't show up for work either Monday or Tuesday, even though my boss called and called. I had no explanation for that either.

I'd always suffered occasional migraines, but now my head was exploding in excruciating pain far beyond anything I'd ever experienced. Even the act of speaking vibrated inside my head, slamming the pain against my skull. It was like being at the bottom of a deep pool of Jell-O, trying to communicate, swimming against the pain. If I tried to walk, each footstep rattled my head like a sledgehammer. Even if I lay still, the pounding was so intense that my teeth clacked together with each beat of my pulse.

I began leaving Wesley piles of mice in case I was suddenly out of commission. They keep for a few days, since they're insulated in fur. At work, I had achieved enough seniority that I was treated with respect and given incredible leeway as I tried to cope with my bizarre condition. I didn't want to admit that anything serious was wrong with me. But it finally got to the point where I couldn't function in the workplace anymore and my doctors and I realized that this was not going to go away anytime soon. So I had to tell my bosses that I was too sick to work, and I went on disability.

Since I lived alone, I had to try to carry on with the basics of life. One afternoon I took some Advil and went shopping for groceries. I loaded the car, sat down in the driver seat, and passed out. I lay there for over eight hours. People in the parking lot eventually noticed I was unconscious and banged on the driver's door. Unable to rouse me, they finally called the police and told them that someone had died in an SUV at the grocery store.

The paramedics broke into my truck and pulled me out. They put me into an ambulance, started IVs, and revived me, but my vitals were bad. When I woke up in the hospital I kept insisting that they let me go home, as it was "no big deal." The doctors insisted that it was a very big deal.

I endured an endless battery of tests, going to this famous clinic and that famous hospital—with the final conclusion being that I had a brain tumor that was inoperable but not cancerous. I had something called massive basilar hemiplegic migraine disorder with complications. The complications included vasospasmic strokes, comas, seizures, ministrokes that cause temporary paralysis, fugue states, narcolepsy, and syncope. It never stopped. The migraine was almost continuous, and I was in excruciating pain 24/7 as they tried this medicine and that treatment, all with bewildering side effects.

If you let some kinds of migraines progress too far you can have a full-blown stroke, as I found out the hard way. Waiting for a prescription at the pharmacy, I decided to try their free blood pressure measuring machine. It read 278/195. I stood up to tell the pharmacists that their machine was broken and collapsed, unable to speak. The same paramedics who had rescued me at the grocery store were rushed over to the pharmacy.

I was in a wheelchair for a month and experienced speech and memory lapses, but eventually recovered. That's the "hemiplegic" part—becoming paralyzed on one side of the body because of the intensity of the migraine, which caused the blood vessels to close, preventing blood from getting to parts of the brain. The tumor, however, eventually calcified and quit growing, but I still struggled with constant excruciating pain.

Whenever my pain level got out of control, I landed in the hospital and they had to give me 20 to 24 units of morphine. To put this in perspective, a badly wounded soldier in the battlefield is given 10 units to treat his pain. Twenty units will kill a grown man. Only a doctor would administer my injection, since no nurse would take that risk. Yet, after this massive dose, my blood pressure fell only slightly. When you're in that much pain, your body goes into a hyper state of "fight or flight," pumped up with so much adrenaline that it counters the efficacy of the morphine. This fight-or-flight state can be so powerful that it accounts for the stories we hear about people possessing superhuman strength in an emergency—a small woman lifting a car up off her child, for example—it overrides your system's normal limits. When I was in this state, my pain could only be dulled.

Another complication from my brain tumor was that I couldn't stay awake for any length of time, a form of narcolepsy. I'd be eating dinner and wake up a half hour later with my face planted in the food or fall asleep standing in the aisle of a store.

I was *really* sick and hardly able to care for myself, much less Wesley. But Wesley's well-being was something worth fighting for, and it focused my attention away from my troubles. I remembered something that Dr. Penfield had told me when I first adopted Wesley, "To that which you have tamed, you owe your life."

The neurologists told me I needed to move closer to a relative, so someone could check on me regularly. Once they explained the seriousness of my prognosis and that I could die, my mother intervened and insisted that I move back to her house.

Wesley and I had lived quite happily with my mom when I first worked in the aerospace industry, but this time around I felt helpless being so dependent and having to ask her for money. My disability insurance company discontinued payment after a while, and I didn't have the resources or energy to take them to court. Paying for all the specialists had emptied out my 401(k) and savings, and I was left with only a tiny fixed disability income—and huge medical bills. It was a nightmare that only seemed to get worse.

All I could think about was that I'd become a terrible drain on my hardworking mother. She and her boyfriend Wally were seriously involved now, and my presence was interfering with any plans they might have to get married. Truly, I'd had a great, exciting life and couldn't have asked for more. Now I was going to be an incurable, constant burden on my mother—financially, emotionally, and socially. Of course, she never said any of these things to me, but there were some panicky times as she looked over her finances and questioned whether she would be able to retire. (My dad had already retired and was dealing with his own health issues.)

My doctors said I would never get better, that I would always be in this state of outrageous pain and lassitude, lost in this deep pit, unable to climb out. I felt like such a medical anomaly; neither side of my family had any history of serious illness. Even in old age, no one had ever been a "burden," but had lived a long healthy life. For three generations both sides of my family were professional musicians or TV personalities. "The show must go on" was our motto.

Even as a child of six, when I had worked in the recording and television industry with my younger sister, we learned to ignore sickness, hiding it from Mom when we had a touch of the stomach flu, for instance. We'd just tough it out, knowing that we would eventually get over whatever illness came our way. But now, here I was, unable to snap out of it, unable to perform the simplest tasks, drowning in pain, breaking all the family rules.

At my lowest point I considered suicide. There seemed to be no other answer. Normally, I think suicide is the most selfish thing a person can do, leaving family and friends in grief and guilt and recriminating what-ifs. Two longtime friends had committed suicide, so I knew firsthand the aftermath for the survivors. But I thought my situation was different. My young friends had had their whole lives ahead of them, but according to the prognosis, my life really was over. I reasoned that, if I thoroughly explained in a letter to my friends and family about how I had lived a happy, satisfying life and didn't want to be an endless burden, then perhaps everyone would eventually come to understand that it was for the best.

But Wesley still needed me. He loved me and would die of shock and grief. Wesley couldn't read a letter or comprehend my explanations; he'd just know that I had abandoned him in his final days. And what would happen to this glorious little soul in his angelic body? Would he end up being handled roughly by strangers who couldn't have cared less who he was? Would the absolute tranquility in those dark inscrutable eyes turn to fear, confusion, pain, knowing he had been betrayed at the end? When would that look of peaceful confidence turn to realization and primal terror?

I *knew* who Wesley was, and I did consider an alternative—that we could die together. But there was no way I could have killed Wesley. That would also have been a betrayal that was just beyond me. I had chosen to tame him and thereby made him vulnerable; I had taught him to trust me implicitly, no matter what. After so many years this trust was perfect and unbroken. I had no right to break it. That would have been an obscene act.

Wesley had been my constant companion, my teacher, and my friend. I now made the decision to honor this little body with the huge soul, and to see him through to the end. I had promises to keep. It was the one thing that I could still do. It's the Way of the Owl. You commit for life, you finish what you start, you give your unconditional love, and that is enough. I looked into the eyes of the owl, found the way of God there, and decided to live.

16

The End

AFTER TRYING EVERY single medication protocol in the latest and not-so-latest literature, my neurologists hit upon a combination of prescription drugs that worked to mitigate the pain. Even better, the meds didn't make me "high" and didn't make me gain as much weight as the other medications. From 2000 to 2002 a new doctor was finally able to get me off the roller coaster of hospital stays and drug treatment experimentations. My mother found me a place to live just a few blocks from her house and helped me move in. Wesley came along with me, of course, and I installed him on his perch in my bedroom.

I was still very sick and slept constantly. I fed Wesley and we talked, but we hardly ever cuddled anymore. He mostly slept and I mostly slept. "Teep deep deep, teeple teeple teeple . . ." We slept for a couple of years, he and I. Wesley was eighteen years old, ancient for a barn owl, so he didn't seem to mind.

One day I looked into his obsidian eyes and saw a faint grayness. Very faint, but as time went by, there was more and more. Wesley was going blind. But he didn't show any signs of distress; he still looked healthy, beautiful, and glossy. He did finally lose interest in mating on my arm, however, after one last halfhearted attempt. And that July, he skipped his molt. I waited and waited, wondering if perhaps I had altered the light cycle that triggers the molt by putting a night-light in my bedroom.

Wesley was also getting stiff; his movements now were slow and tentative. His poop stuck to the feathers around his bottom. He could no longer reach back there to groom, so I trimmed his feathers and, like a nursemaid, cleaned that area for him every day. He began to screech at odd hours of the night, and I had to move into another bedroom to get enough sleep.

His sight was failing, but his hearing was still sharp. If I walked quietly downstairs or got up from the couch, Wesley could hear my footfall, even barefoot on carpet, even with his door closed upstairs. He would screech in acknowledgment and I would answer "Hi, baby!" Although I could no longer sleep in the same room because of his erratic screeching, he was always aware of my presence, and we conversed back and forth. I could speak to him in a soft voice from the opposite side of the town house and he could hear me perfectly.

I kept thinking Wesley would molt soon because he was demanding more and more mice. And more. However, I'd find whole mice and mouse parts left on the perch despite his begging for food (a very distinct verbalization). Perhaps he could no longer see the mice, though he certainly could feel them with his feet. Was he just getting picky? I would give him new mice, which he would attack as if he'd been starving for days, sometimes holding one mouse in his beak and clutching another in his talons. He would turn his back to me, and hunch over the mice protectively. It was strange. Why was he so paranoid and worked up about food? Did he have some eating disorder? I usually could figure out what he was trying to communicate, but now I was confused.

One day Wesley was hanging upside down off the edge of his perch, as he loved to do. He'd hang there by one leg, perfectly still, and quietly observe the world. After tiring of this amusement, he'd grab the towels on the perch with his talons and chin-up onto the platform. By habit, I'd walk by his perch and cup my hand under his hanging body, and swing him back up onto his platform. But when I did that this time, he fell forward off the perch and just hung there by his leash. I tried again. He fell. I unhooked Wesley and set him on the floor and he fell forward on his face. What was going on here?

I decided his eating problem had gone too far and he was probably hungry or dehydrated, so I injected him with a large dose of IV fluid with electrolytes and held him in my arms until his blood sugar rose. My diagnosis was correct and he was fine. But now I knew he couldn't stay on his perch anymore. He wasn't eating his mice properly, and if weakened, he could no longer pull himself back onto it.

It was time for new quarters. I had a super-size travel carrier—the kind used for very large dogs—for Wesley, which he adored. It was exactly the kind of nesting site he would have selected as a young owl, with lots of room to flap his wings and walk around. I set him up with dark towels draped over a stack of books set on the carrier floor, so he had a soft, elevated perch. I filled a heavy white dish with seven or eight mice per day. This new living arrangement made it easier for him to find his mice. And now that he knew he always had mice, he settled down and no longer begged for food.

Things seemed to improve for a while. I still believed that I could handle whatever new situation Wesley threw at me, since for almost twenty years I had always been able to improvise with very little outside help. But not long thereafter, Wesley began to beg for mice again, even when he knew they were there.

This drove me to tears. Everything else about him seemed fine, his feathers were gorgeous, he still played in shallow water and took face baths and drank, so he was generally well hydrated, even if he didn't eat the whole mouse. But he was too skinny.

One day I was driving home from an appointment and took a country road, the scenic route. I was listening to a morbid Scottish song, translated into Irish Gaelic, about a man discovering his beloved—dead and frozen to her bed during the famine. I was always learning songs in Irish, but this one was even more morbid than most. As I drove along, singing this lament, I suddenly spotted a dead barn owl by the side of the road. My heart leapt into my throat. It was like seeing my own child, my Wesley, dead by the road. I stopped to check, just in case it was alive and not moving. This gorgeous young owl was unmarked but dead. I had a horrible premonition and felt ice cold as I stood in the warm sunlight. I picked up the owl, cradled it, and took it out to the field nearby. I got a camp shovel out of the truck and buried him slowly and carefully, knowing now that this was a portent. Tears streaming down my face, I kept saying to myself, "*No*, I can't lose him. I just can't."

The one thing I hate about animal stories is that after you've almost read the entire book and you really care about the animal, they go and tell you all about how the animal died. In fact, I often read the end of these books first so I can at least brace myself for the inevitable. So you should stop reading now if you don't want to hear about Wesley dying. But I need to tell you.

Of course, I knew it would come. Wesley was ridiculously old. Dr. Coward said it was like a human still living at 120. He told me I had taken exceptional care of Wesley, there wasn't a single stress line in his feathering; he was perfectly groomed and gorgeous. When I protested about how skinny he was Dr. Coward said, "Well, a 120-year-old man is very skinny and weak, too, you know."

If only I had cuddled him more that last year, but he didn't seem to want to be cuddled. If only I had spent every last moment with him instead of sleeping so much on the couch. If only I had figured out what was going on with the mice. But there's nothing I could have done. I had done everything I knew for Wesley and I knew a lot, but Wesley was masking an illness from me, as wild animals do.

The night of January 8, 2004, at a time when Wes would normally wake and start talking to me, I heard a weird sound upstairs. I immediately raced up to his room, knowing something was very wrong. Wesley was in his carrier house, trying to greet me, but the only sound he could make was a raspy wheeze. I brought him out of his carrier and stood him on the floor. He swayed. Oh, no. Please don't sway. He made a pathetic whistling noise. What was wrong? Did he need water? I gave Wesley water and he took a sip but couldn't swallow properly and it came back out of his nose. I dried him off and began feeling desperate.

Wesley was so weak it must mean he needed nutrition. I felt his tummy and it was empty. I brought him a mouse, but he wouldn't even look at it, so I tried to force-feed him. I fed starving raptors as a wildlife rescue worker, so I knew what I was doing. After a few minutes, he spat out the piece.

I held Wesley in my arms and raced downstairs to my animal medical station and gave him IV fluids. He lay on his back on the couch, gazing at me with his solemn eyes. He didn't seem perturbed. I kept telling him he was okay, I was with him, I loved him, I'd help him. I was sobbing now. I called Dr. Coward's office and cried, "My bird is in an emergency! He is collapsing! Tell Dr. Coward!" They replied, "Come right away. Don't worry that it's after closing time. We'll be waiting for you."

I tucked Wes into his small travel carrier and propped him

up with little pillows and rolled towels so that he wouldn't have to brace himself during the drive. He lay his head down and watched me silently as I put the seat belt around his carrier and jumped into my SUV. I opened his carrier door and caressed his head and talked to him all the way to the vet.

"It's okay, Wes. I love you. I'm your mommy, I'm here. It'll be okay. Just rest. Dr. Coward will help you. I love you so much. You are my light and my joy. You've been my best friend and I'll always love you. Thank you for letting me into your life. Mommy loves you so much."

I ran into the vet's office with Wesley in his carrier, then gently lifted him out and cradled him in my arms. Dr. Coward gave him more fluids and a shot of vitamins and tried some other tricks of the trade. Wesley didn't even try to resist. By now we were all crying in the room, and I was spilling out my guilt about why didn't I do more and what if I could have done this or that for him or figured it out . . . And Dr. Coward kept telling me it was a miracle that Wesley had lived this long, that my care for him was possibly the best he'd ever seen as a vet, that Wesley obviously had an incredibly full and happy life and that he had grown older than was seemingly possible.

I cradled Wesley in my arms as I had done for the last nineteen years. Dr. Coward brought something over to give Wesley and as I lifted him to the table his head fell forward onto his chest. I held his head up but it fell again. I looked into his eyes. He seemed far away. Was he dead? Dr. Coward grabbed his stethoscope, felt around . . . yes. Wesley was gone. Everyone was crying. One of the vet techs threw herself on me and kissed me on the cheek. I was wet with tears.

Dr. Coward slipped out of the room with Wesley and did a very small, informal exam to see if he could tell me exactly what he had died of. It was massive liver cancer.

He told me, "I don't know how he was alive at all. There's no viable tissue left. It's all cancerous. All of it. You did absolutely everything humanly possible for him, but there was nothing you could have done about this."

And I *had* done the best I could to make Wesley comfortable. I had protected him, kept him warm, fed him, made him feel safe, and we made it to the end without anyone hurting him. We'd made it. All of my prayers for Wesley had been answered.

Wesley changed my life. He was my teacher, my companion, my child, my playmate, my reminder of God. Sometimes I even wondered if he was actually an angel who had been sent to live with me and help me through all the alone times. He comforted me; many times I cried into his feathers and told him my troubles and he tried to understand. He listened and cuddled with me.

He chose to sit on my pillow while I napped and he washed his face when I washed mine. He tried to feed me his mice and make me his mate. He created hundreds of nests for me. He joyfully poured out his love in loud exclamations and had boisterous opinions about everything. He kept a running commentary on all that happened in our lives, in his owl language. He brought us wild owls to the bedroom window with his joyful and jubilant sounds . . . We were happy together.

I'm sorry I couldn't do more for Wesley at the end. I *did* take good care of him and I loved him completely. He was amazing, curious, joyful, strong willed, full of life, a huge soul. His eyes were indescribable. I saw eternity in them, and now at last he was free to fly. My last prayer is that we be reunited in the afterlife, and that he is with God now and that God is taking care of him.

When he died Wesley was lying across my left arm like he always did, with his talons dangling down, his head in my left hand, his eyes closed, and I was grooming him with my right hand. That's when his spirit went out of his body. I'm glad we had that one final embrace.

17

After

A relaxed and sleepy three-year-old. *Stacey O'Brien.*

AFTER WESLEY'S DEATH I fell into a stupor. I hardly slept, instead pouring out my grief by writing our story day and night, driven by a passionate need to remember. My mom graciously stepped aside while I monopolized her computer, even though this virtually shut down her real estate business. In three weeks I wrote the rough draft for this book. Then I slept for months, in and out of a fog.

Before I got sick, I had been taking Irish fiddle lessons from Cáit Reed, a woman who is a top Irish fiddle player in the United States. She plays in the subtle East Clare style that I love. My illness made it impossible to continue lessons, but by then Cáit had become one of my dearest friends. She was the first person to recognize that I was truly sick and needed help, although I was still trying to hide the seriousness of my situation from friends and family. She stepped in early and often took care of me.

A few months after Wesley died, Cáit invited me to a writers' group in Palos Verdes. She'd often bring me to her house so I could sleep until the last minute, then would drive me to the meeting to read my story about Wesley to the group. The first time I went to the writers' group, I walked by a shop with a stuffed barn owl in the window, which I took as a sign of encouragement and bought. A few months later, when I used the library conference room to work further on the book, a great horned owl came and sat in a tree just outside the room for the entire day while I worked. I kept going outside to check if he was really there and I wasn't imagining it.

As the months dragged on, something changed: I began to recover. Since my prognosis had been hopeless, I didn't even notice at first what was happening. But when I compared my current condition to six months before, I could tell there was improvement. I had switched health care providers and my new doctors at Kaiser Permanente found a way to control my symptoms so I was more likely to have days when I could actually function. They continued to tweak my treatments and didn't just throw medications at the problem. My primary physician, Dr. Felder, a man of great intellectual curiosity who worked like a research scientist, always went the extra mile. I learned from this experience never to lose hope and never to take a bad prognosis at face value.

I found solace in continuing to volunteer, when I was able, at wildlife rescue and rehabilitation centers, particularly with wetland and seabirds, birds of prey, and possums. And wild owls still visit me. Even as I wrote these words a barn owl flew over my bedroom and screeched as he went off for his evening hunt. I feel connected to Wesley through these owls. He saved my life; they help me keep going.

My two friends, Cáit and Wendy, held me up emotionally during this time, kindly calling me every day as I wrote and rewrote this memoir. I had been sick and shut in for so long, hardly ever leaving the house, that I was isolated. Their friendship was the lifeline that got me out into the world despite my poor health and losing Wesley.

Wendy remarried and has been very happy for many years. She moved to Colorado and started raising Friesian, Andalusian, and Warlander horses and Ragdoll cats. An award-winning painter and sculptor, Wendy has also been a successful recording artist for decades. She also edited this book with me, sometimes going all night until we both fell asleep on the phone, helping me remember things about Wesley and his extraordinary life. Her husband, Don Francisco, is also an acclaimed recording artist and the kindest, gentlest man a woman could ever hope to find. Annie is now a grown married woman, still wise beyond her years, with a recording career of her own. Oddly enough, Cáit and her husband, Richard Gee, also moved to a high-altitude mountain paradise in Colorado, and they get together with Don and Wendy to make music.

I find great joy in talking about Wesley and sharing his life story with other people. Although I still sometimes feel guilt about his final days, I now know that this is a normal part of grief. Guilt is just anger turned inward—anger at our helplessness in being unable to change the inevitable. But we are not gods. We outlive our animals. There's no way around this. So we choose whether or not to take the pain with the joy. I know people who have decided it's too hard and have given up living with animals. But to me there's no question that it's all worth it.

My sister and I had made a vow when I was eight years old. We would live our lives not by staying in the shallow, safer waters, but by wading as deep into the river of life as possible, no matter how dangerous the current. We knew that we had only one chance at this life and we decided to try to make every moment matter. It may seem an odd vow for two little girls to make, but considering the intensity of our childhood—working almost full-time in the recording industry and also spending so much time at Caltech—we had had the unusual opportunity to see many different kinds of people living very different lives. Both of us have kept this vow. Neither of us regrets living this way.

Wesley taught me the Way of the Owl. In the human world, your value as a person is often intrinsically linked to your wealth or most recent accomplishment. But all the accoutrements of the material world were stripped away from me when I got sick. Wesley made me realize that if all I had to give was love, that was enough. I didn't need money, status, accomplishment, glamour, or many of the empty things we so value.

As much as I still mourn him four years later, there's nothing I'd like more than to adopt another baby owl, to take what I learned from Wesley to the next level. This time I would document and record every little thing—each verbal adaptation, each change in vocalization, every instance of his learning my language. I would make my observations official so they could be verified and stand up to scientific scrutiny.

We are on the cusp of a new understanding of animal communication. It was recently discovered that ravens solve problems by thinking them through logically, without need for trial and error. Ravens not only use tools, they create tools, making modifications that rival those of the great apes. Alex the famous African gray parrot, who passed away as the final chapters of this book were being written, exhibited an astounding level of intelligence and proved that he had truly acquired language—he understood what he was saying and what was being said to him. Able to create new word combinations to describe objects he had not encountered before, Alex was on the verge of proving even more complex abilities of symbolic thought when he died. May he rest in peace and may he be remembered as a pioneer on this journey of exploration into the intelligence and sentience of the creatures with whom we share the earth.

There is so much more to be discovered, and I'm sure in decades to come we will look back at this time as one in which we were emerging from the dark ages of understanding animals, their intelligence, and their emotional lives.

My life was forever changed by a single barn owl named Wesley. I will always be grateful to him for teaching me the Way of the Owl.

Some Things You Might Not Know About Barn Owls

1. Barn owls are extremely emotional and have many ways to express their feelings. They are affectionate. They cuddle with their mate and babies. In fact, the best enrichment for an unreleasable, captive owl is affection. Unreleasable barn owls who are not touched will become neurotic and depressed.

2. Barn owls mate for life. If a mate dies, the survivor often shuts down and wills itself to die—much faster than he or she would have died of starvation or dehydration.

3. Barn owls practice birth control. They will breed only when they perceive an excess of available mice for food. In a facility where I worked, we found that when barn owls were given three daily mice each, they left some of the food uneaten, and they bred. When we lowered their portion to two daily mice, they stopped breeding.

4. When barn owls are seen mating, they are often not actually breeding. They use the mating ritual as a greeting—as do many species. Humans use aspects of their mating behavior as a greeting, as in hugging, kissing, and touching.

5. Barn owls can lay one to ten eggs when breeding, although

they most often lay five. They lay one per day, incubating them from the beginning so that the babies hatch in the same order, one per day. This means that the oldest will be bigger and stronger than the next oldest, and so on. In a good year, all the babies will get fed. But in a bad year, the older and more aggressive babies will get the greatest share of the food and live, while the younger and weaker ones will die of starvation. This is the natural course in the wild: it's better for one or two babies to survive than for them all to die of malnutrition over time.

6. Barn owls can "make faces," unlike most birds, because they have many tiny facial muscles. They use facial expressions to communicate, as humans do.

7. Barn owls use many vocalizations to communicate with their mate or keeper, from screams and hisses to snores and chirps to rapid staccato calls and parrotlike squawks. They can vary their original sounds to indicate their emotional state. Unlike songbirds, barn owls change their vocalizations according to the experience they're having at the time.

8. Owls make distinctive side-to-side head movements when investigating something visually, in order to have depth perception. Humans have moveable eyes. Our eyes move rapidly from side to side constantly, as our brains calculate the difference in perceived movement from one object to another. This difference determines the perceived depth of one object compared with another. Because owls' eyes are fixed in their heads, they have to move their entire head in a side-to-side saccade for their brains to obtain this data.

9. Barn owls have a comb on the inside of the talon on the

middle toe. They use it to comb feathers on their head and face, then they clean the comb with their beak.

10. Most owls have an oil gland at the base of their tail called the uropygial gland. They pinch this gland while grooming and spread the contents all over their bodies. This oil is a precursor to vitamin D and, when exposed to the sun, becomes vitamin D, which the owls then ingest when preening. Barn owls have this gland but there is no oil in it. In order to get vitamin D, they swallow their prey whole, ingesting the digestive tract of the mouse.

11. Barn owls are solitary adults, except when with a mate. They do not have a flocking instinct. Keepers of captive owls should not try to use training or domestication or correcting techniques that work with social animals—animals that flock together or live in herds or packs. Social animals understand correcting techniques, which include quick movements or yelling, but these scare barn owls and make them fear the person who makes them. From then on, the barn owl remembers that person as someone who threatened him, and will become more and more terrified, frantic, and uncooperative. The keeper will also become more frustrated, thinking, *I've worked with animals for 30 years. What is wrong with this owl? It must be stupid.* Owls are far from stupid.

12. Owls do not have a crop like other birds do, so they cannot carry more food than they can eat at one time. Therefore they must resort to caching food for the future and feeding on it when they've digested the food in their stomach. They are sometimes seen holding their next meal in their foot rather than caching it where it could get taken by another animal. A

male barn owl trying to attract a mate will pile mice up in a potential nest and try to lure the female with mouse gifts.

13. Owls are nocturnal and hunt at night. Their most important hunting hour is the first hour after sundown, so they should especially not be disturbed for any reason during this time. Owl watchers should time their walks for after midnight to give the owls time to acquire enough food for the night. Also, humans should avoid even touching the trunk of a tree containing an owl: raccoons, curious about the human scent, will climb the tree and find and kill the owl. Playing tapes of owl calls can interfere with owls' ability to breed and raise young, sometimes entirely stopping all breeding in an area.

14. Owls fly completely silently and have specialized feathering to allow this. They are not waterproof, because their feathers are more like down than like regular bird feathers. The flight feathers have a velvety substance on the surface that absorbs sound, and the edges are serrated/feathered so that there is no distinct edge to swish through the air and alert their prey. The silent flying also allows the owl to use his specialized hearing abilities accurately without interference from the sound of his own feathers.

In part because of their absolute silence, owls sometimes surprise people, leading to stories of ghosts in the woods. And the scream of a barn owl is so terrifying that it has led to stories of hauntings in abandoned buildings and banshees in the night. (On a popular TV series called *Ghost Hunters,* the screech of a barn owl led one researcher to say, "That sounds exactly like a dragon." I wonder how he knows what a dragon sounds like . . .)

15. Barn owls can capture prey in absolute darkness using only their acute sense of hearing. They do not use sonar! Their specialized brain cortex processes auditory information so accurately that the owl has an auditory map of its surroundings, just as humans have a visual map of our world in our minds. Owls can accurately strike a mouse under 3 feet of snow based upon the sound of the heartbeat. Barn owls can triangulate the location of prey in the darkness, because their ears are asymmetrical: The right ear is located high on the head and the left is low on the head, with the right ear pointing down and the left ear pointing up. In humans, the brain calculates the infinitesimally tiny difference between when a sound reaches one ear and when it reaches the other, using the calculation to pinpoint the sound's origin in space along a horizontal plane. Because human ears are symmetrical, we have to cock our heads to add the vertical information. Barn owls' brains simultaneously calculate the difference between when the sound reaches one ear and another along the vertical and horizontal planes to find the sound's origin in three dimensions instantly.

The barn owl brain also calculates the difference in intensity from one ear to the other, which the human brain does not do. Because of the shape of its earflaps, the barn owl can also hear sound from behind in the same way that humans would if we turned our ears backward. Its cuplike facial disks act like a satellite dish, collecting and focusing the sound toward the ears. Even the facial feathers help the owl calculate exactly where a sound originates. If a barn owl is missing a patch of feathers on his face he will either overstrike or undershoot his intended prey target.

16. Barn owls can recognize themselves in the mirror, a feat usually reserved for great apes and humans. Wesley often admired

his feathers in the mirror after a bath, and would relate to me through the mirror in the way that humans do when they are both looking into the mirror at each other and talking.

17. Barn owls have shown an amazing ability to adapt to new situations and can be found living in cities and suburbs. They roost in abandoned buildings, palm trees, and architectural features of buildings. There is a program in the San Francisco area where barn owl nesting boxes are put up in areas of rodent infestation as an alternative to using poisons. As long as barn owls have an open space in which to hunt, they will take up residence in the boxes and solve the rodent problem. To find the best barn owl nest box instructions, refer to www.hungryowl.org.

18. Among the main killers of barn owls are cars and trucks—owls hunt by flying at about the same height as a car or truck windshield, so they are often hit by vehicles—and rodent poisons. The owl eats the poisoned rodent and bleeds to death internally or dies of seizures. Poison meant for rodents also kills hawks, cats, dogs, and other local wildlife.

Acknowledgments

It is no exaggeration when I say that there are some people to whom I owe a lifetime debt of gratitude. When I undertook this project I was still quite ill, and many people made it possible for me to continue forward with their generosity of spirit and kindness, going the extra mile. Without the continued support and effort of Wendy Francisco, I doubt that I would have been able to find an agent, revise the manuscript, get out of bed to write, work through times when I felt blocked, pay expenses related to the book, or edit it. She paid for me to go to the conference where I met my agent, did the artwork and photography and Web site (which she also bought) on time for the conference (she is still the webmaster and creator), encouraged me on the phone every day, paid many of my other expenses, and finally, edited every line with me before I submitted my manuscript to Free Press. Wendy is also the photographer of both pictures on the jacket of this book. She was a relentless and passionate creative partner. We worked together, both of us on headsets, on the phone between Colorado and California, for long hours going over and over the book. It was my first book and her first editing project, so we discussed every aspect of the book until we were, at times, both falling asleep on the phone. We laughed, cried, and relived every moment of this memoir together. After all, it was Wendy, many years ago, who enthusiastically took Wesley into her home as a tiny baby owlet, back when we were roommates. She once delivered kittens while on the phone editing the book with me.

Cáit Reed is another dear friend without whom this book would not have been written, and to whom I owe a lifetime debt of gratitude. Cáit brought me to her house to stay and brought me to her writers' group, insisting that I read my rough draft aloud to them. They became a fundamental sounding board, and I learned about the industry and craft from them. Cáit also called me every day and said she would "not allow" the project to falter. She listened, cajoled, laughed with me, and cried with me over Wesley and his book. She was also indispensable as an objective adviser and copy editor. When I worked on the last draft, I stayed in her home and she helped me with it.

I was encouraged constantly by my family: Ann (my mother) and Wally Farris, Hack (my father) and Emily O'Brien, Warren and Roberta O'Brien, Alicia O'Brien and Tom Kramer, Janis Truex, Linda Caughran, Karen Sandoval, Carol Gilpin, Gloria Gherhart, my sister Gloria O'Brien Fontenot, Ann Blumberg, my Grandpa Haskell (Hack) O'Brien Sr., Grandma Zimmie O'Brien, and by my friends: Cat Spydell; Brenda Gant; Keith Malone; Arleta Okerson; Don Francisco; the Countess Erin Gravina; Linda Conti; Dr. Chandler; Ruth Vollert; Elizabeth McGrail; Mr. Ehring; Lynne Hannah; MyoMyo San; Mike O'Brien; Henry Law; Aileen, Vern, and Guy Ritter; and many others whose enthusiasm touched my heart and kept me going.

My parents gave me a passionate love of reading and writing from the beginning, and passed on their equally passionate love and understanding of animals. They have been a constant encouragement all my life, never doubting that I could do what I set out to do. My mother has helped me and supported me since I became disabled and is still helping me keep my head above water.

Richard Gee, Cáit's husband and my lawyer, advised me many times and helped me with all contracts and legal questions.

I want to thank the Southwest Manuscripters Writers' Group that meets in the Palos Verdes Peninsula Library for teaching me all they knew about writing and the industry, what to expect, and how to make my way through the process of finding publication. In particular I want to thank Jean Shriver for convincing me that I should seek publication, Cat Spydell for pulling the fat out of the fire when I had no idea how to write a book proposal. She set up shop in the library with me, we put all my notes together, and she pulled together the basis for the proposal that we used, while a great horned owl sat outside the window to keep us company. Cheryl Romo constantly encouraged me and was the one who urged me to sign up for the Southern California Writers Conference.

Again, people who did not need to went the extra mile for me. Michael Steven Gregory took special notice of *Wesley the Owl* and created buzz in the conference before I even got there, as did Wes Albers and Chrissie A. Barnett. This conference taught me so much that it prepared me for the next steps, but most important, it's where I met my agent, Sally Van Haitsma. I don't think there is a more dedicated agent anywhere. She was perfect for a new author and has been more than "just" a literary agent by acting as a coach, explainer, mentor, friend, confidante, and ally. She has a work ethic that would make a Puritan blush and never, ever gave up. I am eternally grateful to her for her dedication to the book and to me and her unwavering belief in the project. She went beyond the call of duty by also helping to edit and smooth out the manuscript that we sent to Free Press, offering advice and input on the writing as well.

I also wish to thank Melinda Roth for her help early on with the structure of the manuscript. She was particularly helpful with the first several chapters' structures and themes.

Wendy was especially insightful with the structure of the last two-thirds of the book, too, and had a lot of good ideas and input. Wendy edited the book with me. She is amazing.

I've heard that most books undergo about twelve rewrites before publication. I never believed that, but now I do. I'm not sure how many times I revised the book with help and advice from my editors, but each time, it and I got better, so I'm grateful to everyone who read the manuscript, advised me, critiqued it, and helped with the editing. My parents and sister also critiqued and helped edit the manuscript.

Thank you to Robert S. Nord and Jany Poitras of Heritage Studios for editing video of Wesley and me, creating DVDs and pictures that are now usable. Also many thanks to Line on Line, Inc. for helping to verify and research some of my questions for accuracy.

I also wish to thank the people who loved and accepted Wesley and made his happy life possible: my mother and sister, Wendy, Cáit and Richard, Guy Ritter (who once drove two hundred miles every day to feed Wesley when I had to leave for a funeral), Deborah Hicks, Rich Buhler, Connie Fossa, Raylene, Gene, Kurt Mastellar, Jim Tenneboe, and Dr. Penfield.

Thank you to those of you who answered my many questions and made yourselves available, often going the extra mile: Terry Mingle at Cornell Laboratory of Ornithology, Dr. Don Kroodsma, Dr. Douglass Coward (Wesley's veterinarian) and his wonderful staff, Dr. Weldy, Dr. Cleland, Dr. Moscovakis, and all the Caltech postdocs from way back when.

Thank you to Kaiser Permanente for putting together an amazing team of doctors to manage my condition and make this all possible: Dr. Rosenberg, Dr. Felder, and everyone at Kaiser who has helped along the way.

For believing in me before I'd ever been published, I wish to thank Ja-lene Clark and Laura Wood at Council Oak Books.

Special thanks to Jane Goodall for being my inspiration since I was eight years old, for daring to go where no one had gone before, both physically and scientifically. Dr. Goodall continues to find that animals are far more sentient and intelligent than we had ever imagined. She broke down the barrier between humans and our fellow animals, and continues to do so.

On a lighter note, I must admit that I would not have gotten through this process without the help of Coca-Cola, Lindt 70% Cacao chocolate, and Irish/Celtic music: the Bothy Band, Paddy Canny, Solas, Altan, and Ashley MacIsaac in particular.

At Free Press, I want to thank editorial assistant Donna Loffredo, who kept everything together, input edits, got all the pictures together, and put in the proper places—she was the Grand Central Station for this project and was always calm and collected. I also want to thank Andrew Paulson, former editorial assistant. Thanks to Jennifer Weidman for her insights. I am also deeply indebted to Dominick Anfuso, Andrew Dodds, Suzanne Donahue, Eric Fuentecilla, Shannon Gallagher, Carisa Hays, Martha Levin, Edith Lewis, Patricia Romanowski, Ellen Sasahara, and the whole team at Free Press. Thanks also to the sales and marketing staff for their behind-the-scenes efforts and for the lovely cover. This book would never have become what it is without the passionate care and dedication of my Free Press editor, Leslie Meredith. She is an empath for animals and authors, and is especially involved and attentive in a way that is very rare and dear. She poured her heart and soul into *Wesley the Owl*, improving it without changing its essence, understanding intuitively where to add, where to take away, sculpting, shaping, and shining it. She always had time for me and my new-author questions and concerns, even introducing me to people with whom I can network for the betterment of animals, which is her great passion. There cannot be a better editor anywhere in the world, truly.

About the Author

STACEY O'BRIEN IS trained as a biologist specializing in wild animal behavior. She graduated from Occidental College with a BS in biology and continued her education at Caltech, where she became involved with owl research.

Stacey continues to work in biology as a wildlife rescuer and rehabilitation expert. She works with a variety of local animals, including the endangered brown pelican, seabirds, possums, and songbirds. She lives in Southern California.